Large Scale Scientific Computation

Publication No. 51
of the Mathematics Research Center
The University of Wisconsin–Madison

Academic Press Rapid Manuscript Reproduction

LARGE SCALE SCIENTIFIC COMPUTATION

Edited by

SEYMOUR V. PARTER
Mathematics Research Center
The University of Wisconsin–Madison
Madison, Wisconsin

Proceedings of a Conference
Conducted by the Mathematics Research Center
The University of Wisconsin–Madison
May 17–19, 1983

1984

ACADEMIC PRESS, INC.
(Harcourt Brace Jovanovich, Publishers)
Orlando San Diego New York London
Toronto Montreal Sydney Tokyo

ACADEMIC PRESS, INC.
Orlando, Florida 32887

United Kingdom Edition published by
ACADEMIC PRESS, INC. (LONDON) LTD.
24/28 Oval Road, London NW1 7DX

Library of Congress Cataloging in Publication Data

Main entry under title:

Large scale scientific computation.

Includes index.
1. Fluid dynamics--Data processing--Congresses.
2. Aerodynamics--Data processing--Congresses. 3. Numerical analysis--Data processing--Congresses. I. Parter, Seymour V. II. University of Wisconsin--Madison. Mathematics Research Center.
QA911.L32 1984 532'.05 84-45237
ISBN 0-12-546080-5 (alk. paper)

PRINTED IN THE UNITED STATES OF AMERICA

84 85 86 87 9 8 7 6 5 4 3 2 1

CONTENTS

SENIOR CONTRIBUTORS

Numbers in parentheses indicate the pages on which the authors' contributions begin.

LOYCE ADAMS (301), *Institute for Computer Applications in Science and Engineering, NASA Langley Research Center, Hampton, Virginia 23665*

JOEL E. DENDY, JR. (1), *Theoretical Division, MS-B284, Los Alamos National Laboratory, Los Alamos, New Mexico 87545*

ROLAND GLOWINSKI (23), *Université Pierre et Marie Curie, 75230 Paris Cedex 05, France and INRIA, 78153 Rocquencourt, France*

JAMES M. HYMAN (51), *Los Alamos National Laboratory, Los Alamos, New Mexico 87545*

AKIRA KASAHARA (85), *National Center for Atmospheric Research, Boulder, Colorado 80307*

H. T. KUNG (127), *Department of Computer Science, Carnegie–Mellon University, Pittsburgh, Pennsylvania 15213*

DENNIS R. LILES (141), *Safety Code Development Group, MS-K553, Los Alamos National Laboratory, Los Alamos, New Mexico 87545*

R. W. MACCORMACK (161), *Department of Aeronautics and Astronautics, FS-10, University of Washington, Seattle, Washington 98195*

C. WAYNE MASTIN (195), *Department of Mathematics and Statistics, Mississippi State University, Mississippi State, Mississippi 39762*

V. L. PETERSON (215), *NASA Ames Research Center, Moffet Field, California 94035*

ROBERT E. SMITH (237), *NASA Langley Research Center, Hampton, Virginia 23665*

PAUL N. SWARZTRAUBER (271), *National Center for Atmospheric Research, Boulder, Colorado 80307*

PREFACE

This volume is the collected invited lectures presented at a Conference on Large Scale Scientific Computation in Madison on May 17–19, 1983, under the auspices of the Mathematics Research Center of the University of Wisconsin, sponsored by the United States Army under Contract No. DAAG29-80-C-0041 and supported in part by the Department of Energy Grant No. DE-FG02-83ER 13074, by the National Aeronautics and Space Administration under Contract No. NASW-3829, and by the National Science Foundation under Grant No. MCS-7927062, Modification 2.

This conference was devoted to a review of the various and diverse developments of modern large scale scientific computation. Speakers at this meeting addressed questions concerned with

1. Specialized architectural considerations
2. Efficient use of existing "state-of-the-art" computers
3. Software developments
4. Large scale projects in diverse disciplines
5. Mathematical approaches to basic algorithmic and computational problems

Thanks are due to Gladys Moran for the careful handling of the conference details and to Joyce Bohling for the handling of all of the many tasks (including nagging the authors and the editor) which went into assembling the volume. Finally, thanks are due to the other members of the organizing committee of the conference. Without their participation, the conference might never have come into being and certainly would have been far less successful. They are Carl de Boor, Mac Hyman, and John Strikwerda.

MULTIGRID SEMI-IMPLICIT HYDRODYNAMICS REVISITED

Joel E. Dendy, Jr.

1. INTRODUCTION.

The multigrid method has for several years been very successful for simple equations like Laplace's equation on a rectangle. For more complicated situations, however, success has been more elusive. Indeed, there are only a few applications in which the multigrid method is now being successfully used in complicated production codes. The one with which we are most familiar is the application by Alcouffe [1] to TTDAMG, stemming from [2], in which, for a set of test problems, TTDAMG ran seven to twenty times less expensively (on a CRAY-1 computer) than its best competitor. This impressive performance, in a field where a factor of two improvement is considered significant, encourages one to attempt the application of the multigrid method in other complicated situations.

The application discussed in this paper was actually attempted several years ago in [4]. In that paper the multigrid method was applied to the pressure iteration in three Eulerian and Lagrangian codes. The application to the Eulerian codes, both incompressible and compressible, was successful, but the application to the Lagrangian code was less so. The reason given for this lack of success in [4] was that the differencing for the pressure

ISBN 0-12-546080-5

equation in the Lagrangian code, SALE, was bad. For example, on a uniform grid with mesh size h, the approximation to the Laplacian Δ in SALE is the skewed Laplacian:

$$\Delta_h^{sk}\, p_{i,j} = \frac{1}{2h^2}(p_{i-1,j-1} + p_{i-1,j+1} + p_{i+1,j-1} + p_{i+1,j+1} - 4p_{i,j}) \quad ; \tag{1.1}$$

why this is a bad differencing is discussed below. In [4] the differencing for the pressure equation was changed so that on a uniform grid, the Laplacian was approximated by the discrete five point Laplacian:

$$\Delta_h\, p_{i,j} = \frac{1}{h^2}(p_{i-1,j} + p_{i+1,j} + p_{i,j-1} + p_{i,j+1} - 4p_{i,j}) \quad ; \tag{1.2}$$

when this change was made, we were able to apply the multigrid method successfully. In this paper, we examine again the application of multigrid to the pressure equation in SALE with the goal of succeeding this time without cheating.

2. MULTIGRID REVIEW.

To explain the difficulty with (1.1), it is helpful to review the multigrid method. Suppose that the equation LU = F is approximated on a grid G^M by

$$L^M U^M = F^M \quad . \tag{2.1}$$

In the simplest form of the multigrid method, one constructs a sequence of grids $G^1, \ldots, G^M$ with corresponding mesh sizes $h_1, \ldots, h_M$, where $h_{i-1} = 2h_i$. One does a fixed number, IM, of relaxation sweeps (Gauss-Seidel, for example) on (2.1) and then drops down to grid G^{M-1} and the equation

$$L^{M-1}V^{M-1} = f^{M-1} \equiv I_M^{M-1}(F^M - L^M V^M) \quad , \tag{2.2}$$

where v^{M-1} is to be the coarse grid approximation to $V^M \equiv U^M - u^M$, where $v^M = u^M$ is the last iterate on grid G^M, and where I_M^{M-1} is an interpolation operator from G^M to G^{M-1}. To solve equation (2.2) approximately, one resorts to recursion, taking ID relaxation sweeps on grid G^k before dropping down to grid G^{k-1}, $M-1 \geq k \geq 2$ and the equation

$$L^{k-1}v^{k-1} = f^{k-1} \equiv I_k^{k-1}(f^k - L^k v^k) \quad . \tag{2.3}$$

When grid G^1 is reached, the equation $L^1 v^1 = f^1$ can either be solved directly or to some precision by iteration and $v^2 \leftarrow v^2 + I_1^2 v^1$ can be performed. Then one does IU relaxation sweeps on grid G^{k-1} before forming $v^k \leftarrow v^k + I_{k-1}^k v^{k-1}$, $3 \leq k \leq M$.

The motivation behind the multigrid method is that, on a given grid G^k, relaxation methods like Gauss-Seidel generally do a fine job of reducing high frequency components of the error but a poor job of reducing low frequency components. More specifically, let the $\theta = (\theta_1, \theta_2)$ Fourier component of the error functions v and $\bar{v}$ before and after a relaxation sweep on G^k be written as

$\bar{v}_{\alpha,\beta} = \bar{A}_\theta e^{i(\theta_1\alpha+\theta_2\beta)}$ and $v_{\alpha,\beta} = A_\theta e^{i(\theta_1\alpha+\theta_2\beta)}$. Suppose that $\mu(\theta) = \left|\frac{\bar{A}_\theta}{A_\theta}\right|$ is appreciably less than one for components θ with $\frac{\pi}{2} \leq |\theta| = \max(|\theta_1|, |\theta_2|) \leq \pi$; then such components can be efficiently reduced by relaxation sweeps on G^k. Components θ with $0 \leq |\theta| \leq \frac{\pi}{2}$ are the ones which can be approximated on G^{k-1}; components θ with $\frac{\pi}{4} \leq |\theta| \leq \frac{\pi}{2}$ on G^k are mapped to components θ with $\frac{\pi}{2} \leq |\theta| \leq \pi$ on G^{k-1} and can be efficiently reduced by relaxation sweeps on G^{k-1}, while components θ with $0 \leq |\theta| \leq \pi$ on G^{k-1} are mapped to components θ with $0 \leq |\theta| \leq \frac{\pi}{2}$ on G^{k-2}, which are the ones which can be approximated on G^{k-2}. Recursion leads to G^1, which is assumed coarse enough either to solve directly or to iterate efficiently.

For the operator (1.2), $\bar{\mu} = \max\{\mu(\theta):\frac{\pi}{2}\leq|\theta|\leq\pi\} = .5$, and the multigrid method performs admirably [3]. However, for the operator (1.1), $\bar{\mu} = 0$ since $\mu(\pi,\pi) = 0$ for Δ_h^{sk}. Thus multigrid for Δ_h^{sk} performs no better for Δ_h^{sk} than simple relaxation. A numerical example of this bad performance is given in Section 4.

3. RAY MULTIGRID METHOD.

The failure of multigrid for Δ_h^{sk} was a source of annoyance to Brandt, which motivated him to derive a cure, the description of which from [8] we now summarize. First, however, we need to be more specific about the choice of L^k, $k < M$, in (2.2) and (2.3). Assuming L^M to be positive definite, rewrite (2.1) as $(L^M)^{\frac{1}{2}}U^M = (L^M)^{-\frac{1}{2}}F^M$. Then given an approximation u^M to U^M, find V^{M-1} so as to minimize

$E(V^{M-1}) = \left\|(L^M)^{\frac{1}{2}}(u^M + I_{M-1}^M V^{M-1}) - (L^M)^{-\frac{1}{2}}F^M\right\|_{G^M,L_2}$. This minimization problem is equivalent to (2.2) if $L^{M-1} = I_M^{M-1}L^M I_{M-1}^M$ and $I_M^{M-1} = (I_{m-1}^M)^*$. A similar result holds by induction for L^k, $k < M-1$.

Let $S = \{\theta: |L^M(\theta)| << |L^M|, |\theta| \leq \pi\}$, where $L^M(\theta)$ is the symbol of L^M (i.e., the function $L(\theta)$ which satisfies $L^M e^{i\theta\cdot x/h} = L^M(\theta)e^{i\theta\cdot x/h}$), $|L^M| = \max_{|\theta|<\pi}|L^M(\theta)|$, and $h = h_M$. The Fourier modes $e^{i\theta\cdot x/h}$, $\theta\varepsilon S$ are the slowly convergent modes in any reasonable local relaxation process. Hence, G^{M-1} should approximate these modes well. Attempt to write $R^M = f^M - L^M u^M$ and $V^M = U^M - u^M$ as

$$R^M(x) = \sum_{s\varepsilon S'} R_s^M(x)e^{is\cdot x/h} \quad , \; R_s^M \text{ smooth} \quad ,$$

$$V^M(x) = \sum_{s\varepsilon S'} V_s^M(x)e^{is\cdot x/h} \quad , \; V_s^M \text{ smooth} \quad ,$$

where $S' \subseteq S$ is a finite set such that if $\theta\varepsilon S$, then there exists $s\varepsilon S'$ such that $e^{i(\theta-s)\cdot x/h}$ is smooth, i.e., such that $\theta - s << 1$. The role of G^{M-1} is to approximate V_s^M, $s \varepsilon S^1$, well, which should be possible since the V_s^M should be smooth after relaxation. For each $s \varepsilon S'$ we want an equation like

$$L_s^{M-1} V_s^{M-1} = I_{m,s}^{M-1} R_s^M \quad ; \tag{3.1}$$

to do this, consider

$$E(\{V_s^{M-1}\}) =$$

$$\left\| (L^M)^{\frac{1}{2}} (u^M + \sum_{s \varepsilon S'} e^{is \cdot x/h} I_{M-1}^M V_s^{M-1}) - (L^M)^{-\frac{1}{2}} F^M \right\|_{G^M, L_2}$$

$$= \left\| (L^M)^{\frac{1}{2}} (\sum_{s \varepsilon S'} e^{is \cdot x/h} I_{M-1}^M V_s^{M-1}) - (L^M)^{-\frac{1}{2}} R^M \right\|_{G^M, L_2}$$

$$\leq \sum_{s \varepsilon S'} \left\| (L^M)^{\frac{1}{2}} e^{is \cdot x/h} I_{M-1}^M V_s^{M-1} - (L^M)^{-\frac{1}{2}} e^{is \cdot x/h} R_s^M \right\|_{G^M, L_2} .$$

Minimization of each term in the last sum leads to (1.6) with $L_s^{M-1} = (I_{M-1,s}^M)^* L^M I_{M-1,s}^M$, $I_{M,s}^{M-1} = (I_{M-1}^M, s)^*$, and $I_{M-1,s}^M = e^{is \cdot x/h} I_{M-1}^M$.

After solution of (3.1), the correction $u^M \leftarrow u^M + \sum_{s \varepsilon S'} I_{M-1,s}^M V_s^{M-1}$ is made. Of course, by recursion (3.1) can be solved approximately for each $s \varepsilon S'$ by relaxation and by construction of a set S_s' consisting of the slowly converging modes for relaxation on (3.1).

4. APPLICATION OF THE RAY MULTIGRID METHOD TO Δ_h^{sk}.

In this section we consider application of the ray method to the equation

$$\begin{aligned} -\Delta u + 10^{-4} u &= F \text{ in } \Omega = (0, .96) \times (0, .96) \\ \frac{\partial u}{\partial \nu} &= 0 \text{ on } \partial\Omega \quad . \end{aligned} \tag{4.1}$$

The reason we consider (4.1) is that it is a model equation for the problem in SALE considered in [4], which had zero Neumann boundary conditions and a lower order term with a small multiple. The discrete approximation we consider is cell-centered as in SALE. At the $i,j^{\underline{th}}$ cell center, the discrete equation we consider is

$$-\Delta_h^{sk} U_{i,j}^M + 10^{-4} U_{i,j}^M = \cos(25\pi(i-2)h)\cos(25\pi(j-2)h) \,,$$

$$2 \le i \le 25, \; 2 \le j \le 25 \quad . \qquad (4.2a)$$

Here $h = h_M = .04$, and the right hand side is chosen to be rich in the (π,π) frequency.

There are at least two possible approximations to the boundary conditions, which we will refer to as the "finite element" and "finite difference" approximation. The "finite element" approximation (so called because it results from using piecewise bilinear elements on quadrilaterals with midpoint quadrature) we illustrate by giving two typical cases: at (2,2), $-\Delta_h^{sk} U_{i,j} = \frac{1}{2h^2}(-U_{i+1,j+1} + U_{i,j})$, and at (2,j), $2 < j < 25$, $-\Delta_h^{sk} U_{i,j} = \frac{1}{2h^2}(-U_{i+1,j+1} - U_{i+1,j-1} + 2U_{i,j})$. The "finite difference" boundary approximation to the boundary condition can be derived by using fictitious cells, writing down a difference approximation to $\frac{\partial u}{\partial \nu} = 0$ (e.g., $\frac{U_{1,j} - U_{2,j}}{h} = 0$, $2 \le j \le 25$) and then eliminating the fictitious cells in terms of the interior cells. Two typical cases are:

at (2,2),

$$-\Delta_h^{sk} U_{i,j} = \frac{1}{2h^2}(-U_{i,j+1} - U_{i+1,j} - U_{i+1,j+1} + 3U_{i,j}) \qquad (4.2b)$$

and at (2,j), $2 < j < 25$,

$$-\Delta_h^{sk} U_{i,j} = \frac{1}{2h^2}(-U_{i,j-1} - U_{i,j+1} - U_{i+1,j+1}$$

$$-U_{i+1,j-1} + 4U_{i,j}) \quad . \qquad (4.2c)$$

For a uniform grid and $-\Delta$, the differencing in SALE reduces to the above "finite difference" approximation on the boundary; hence, it is of more interest to us. Also, the "finite element" boundary approximation annihilates the (π,π) frequency, but the "finite difference" boundary approximation does not. Not annihilating the (π,π)

frequency is a desirable feature since it is a partial cure for the "hourglass" instability that is troublesome in Lagrangian codes like SALE. (In [7], it is claimed that the "finite difference" boundary approach is a total cure for the "hourglass" instability. In fact, for some problems, it is still necessary to smooth the "hourglass" frequency a little; this is referred to as an "alternate node coupler" and is discussed further in Sec. 6.)

Before we can describe the numerical results for (4.1) we must describe the operators I_{k-1}^{k}. We assume G^{k-1} is every other grid point of G^{k}. (See Fig. 1.)

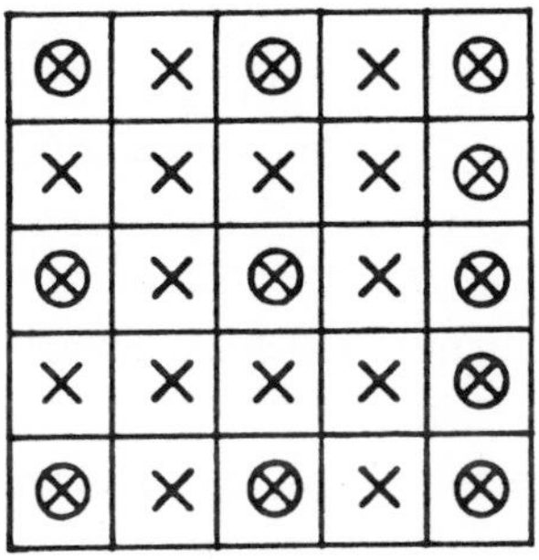

Fig. 1. Two cell-centered grids in which coarse grid unknowns are every other fine grid unknown.

Suppose that $(IF,JF) \varepsilon G^{k}$ is the same point as $(IC,JC) \varepsilon G^{k-1}$. Then at (IF,JF), I_{k-1}^{k} is just given by replacement: $(I_{k-1}^{k} V^{k-1})_{IF,JF} = V_{IC,JC}^{k-1}$. Suppose that at (IF+1,JF), L^{k} is given by the pointwise template

$$\begin{bmatrix} -NW & -N & -NE \\ -W & C & -E \\ -SW & -S & -SE \end{bmatrix} . \tag{4.3}$$

Then

$$(I_{k-1}^{k} v^{k-1})_{IF+1,JF} =$$

$$\frac{(NW+W+SW)v_{IC,JC}^{k-1} + (NE+E+SE)v_{IC+1,JC}^{k-1}}{(C-N-S)} .$$

(We have just summed (4.3) vertically to average out its y-dependence.) A similar formula is used for points like (IF,JF+1). In each case, one performs $v^k \leftarrow v^k + I_{k-1}^k v^{k-1}$. Enough information is now present to use the difference equation at points like (IF+1,JF+1) to solve for $v_{IF+1,JF+1}^k$ in terms of its eight neighbors. Further details are contained in [5]. Since $10^{-4} << 1$, I_{k-1}^k is very nearly bilinear interpolation except near the boundary, where the above formulation gives a good extrapolation for points of $G^k \setminus G^{k-1}$ which do not lie between two points of G^{k-1} or in the center of four G^{k-1} points.

We take M = 4, IM = 2, IU = 1, ID =1, and we use $I_k^{k-1} = (I_{k-1}^k)^*$ and $L^{k-1} = (I_{k-1}^k)^* L^k I_{k-1}^k$. Then the asymptotic convergence factor per multigrid cycle for the multigrid algorithm described in Sec. 2 is .92. By asymptotic convergence factor we mean the ratio of the discrete L_2 norms of the residual on G^4 before and after a multigrid cycle ($G^4 \rightarrow G^3 \rightarrow G^2 \rightarrow G^1 \rightarrow G^2 \rightarrow G^3 \rightarrow G^4$.)

Now let us consider the ray multigrid method for (4.2). The set S is $\{(0,0),(\pi,\pi)\}$. Take $U_1^h \equiv 1$ and $U_{-1}^h = (-1)^{(x+y)/h}$. Define $h = h_M$, $I_{M-1,i}^M = U_i^h I_{M-1}^M$, and $L_i^{M-1} = (I_{M-1,i}^M)^* L^M I_{M-1,i}^M$, $i = -1,1$. Derive interpolation operators $I_{M-2,i}^{M-1}$ as described above from the L_i^{M-1} and define $I_{M-2,i,j}^{M-1} = U_j^h I_{M-2,i}^{M-1}$, $h = h_{M-1}$, $j = -1,1$. Continue recursively. (Thus, there are four $L_{i,j}^{M-2}$'s and eight $L_{i,j,k}^{M-3}$'s.) The asymptotic convergence factor per multigrid cycle for this algorithm is .21. Note that there is 3/2 as much storage and work per cycle for this algorithm versus the regular multigrid algorithm (since $1 + 1/4 + 1/16 + \ldots = 4/3$ and $1 + 2(1/4) + 4(1/16) + \ldots = 2$).

Another algorithm uses the corrections as soon as they are available instead of saving them up. This is done by using W-cycles and the diagram in Fig. 2 for M = 3, should make the algorithm clear. The asymptotic convergence factor for (4.2) per multigrid cycle (now a W-cycle) is .17.

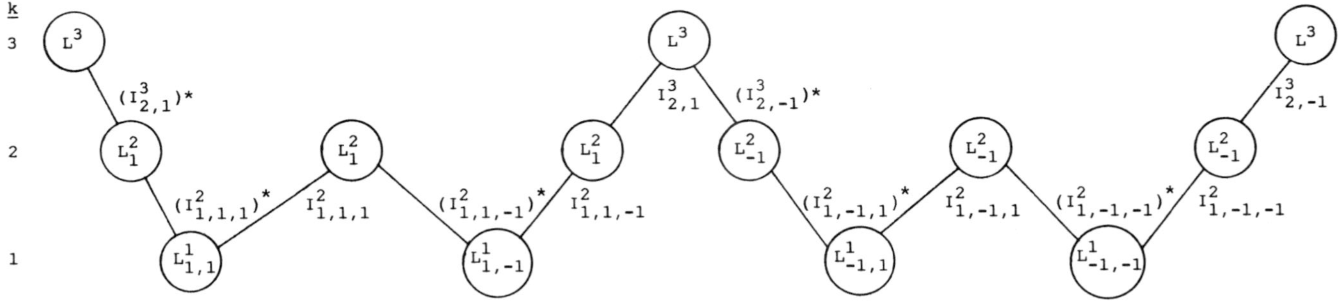

Fig. 2. W-cycle for ray multigrid with three grids.

5. THE RAY MULTIGRID METHOD FOR Δ_h^{sk} WITH ONLY ONE ARM.

As pointed out in [4], a matrix is never explicitly formed in the pressure iteration in SALE. Implementation of either method in the last section requires a matrix to be explicitly formed. What we want to investigate is whether these methods can be modified so as to be applicable without explicitly forming a matrix. As a first step, then, we consider whether L_s^k, $k < M$, $s \in S$, can be formed explicitly instead of from the variational (the $I^T LI$) approach.

First we recall the grid structure that was used in [4]. Instead of forming G^{k-1} by taking every other cell center of G^k, we let G^{k-1} be as in Fig. 3. We take I_{k-1}^k to be bilinear interpolation and $I_k^{k-1} = (I_{k-1}^k)^*$; for points near the boundary of G^k, we use fictitious cells and reflection (to approximate $\frac{\partial U}{\partial \nu} = 0$) to determine I_{k-1}^k; this gives rise to an extrapolation formula near the boundary.

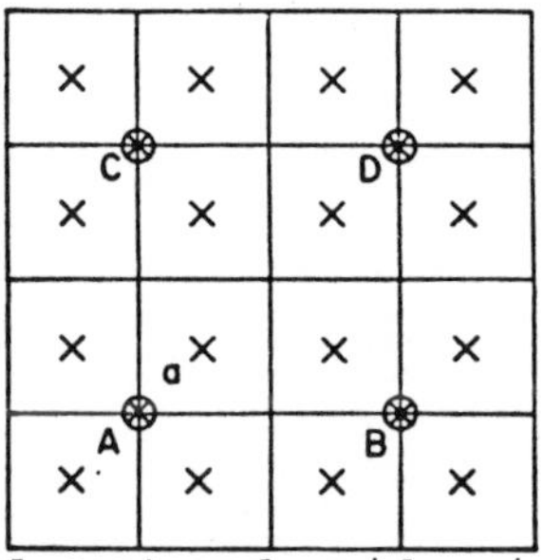

Fig. 3. Two cell-centered grids, in which coarse grids unknowns are not a subset of fine grid unknowns.

Consider first (4.2a) with the "finite element" boundary condition. Following [8], we seek an alternative definition of L_s^{M-1}, $s \in S$ so that

$$e^{is\cdot x/h}\, I_{M-1}^M L_s^{M-1}(e^{i2\theta\cdot x/2h}) \simeq L^M(e^{i(\theta+s)\cdot x/h}) \quad (5.1)$$

for small θ. For "finite element" boundary conditions we claim L^{M-1} can just be taken to be $L^{M-1} = -\Delta_h^{sk}\cdot + 10^{-4}\cdot$, $h = h_{M-1}$. Similarly we take $L^k = L_s^k = -\Delta_h^{sk}\cdot + 10^{-4}\cdot$, $h = h_k$. We take $I_{k-1,i}^k = U_i^h I_{k-1}^k$, $h = h_k$, and $I_{k,i}^{k-1} = (I_{k-1,ik}^k)^*$, $i = -1, 1$, and we use W-cycles. Analagous to Fig. 2, we now have Fig. 4. Also, following [8] we use

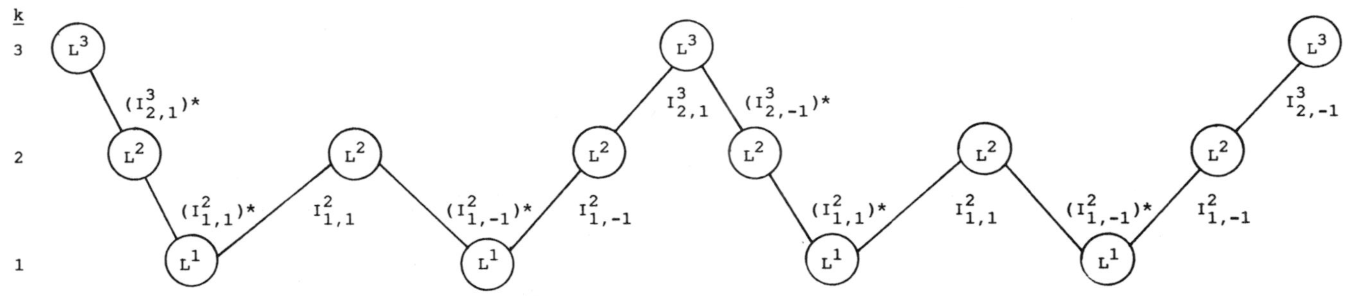

Fig. 4. W-cycle for ray multigrid with operator approximation and "finite element" boundary conditions.

the smoothing operator S^{k-1} to smooth the solution on G^{k-1} before interpolating it to G^k, where

$$S^{k-1} = \frac{1}{8}\begin{bmatrix} 0 & 1 & 0 \\ 1 & 4 & 1 \\ 0 & 1 & 0 \end{bmatrix} ; \tag{5.2}$$

it is not necessary to do such smoothing in the variational approach, but we show by example below that it is necessary in the approaches of this section. We take M = 3, IU = 1, ID = 1, and IM = 1, so that there are two sweeps on G^3 for each W-cycle. The asymptotic convergence factor per W-cycle is 0.20. What should it be? The smoothing factor is

$$\bar{\mu} = \max\left\{ \frac{\frac{1}{2}e^{-i\theta_1}e^{i\theta_2} + \frac{1}{2}e^{i\theta_1}e^{i\theta_2}}{2-\frac{1}{2}e^{-i\theta_1}e^{-i\theta_2} - \frac{1}{2}e^{i\theta_1}e^{-i\theta_2}} , \quad \frac{\pi}{2} \le |\theta| \le \pi \quad , \frac{\pi}{2} \le |\theta-(\pi,\pi)| \le \pi \right\}$$

which is $\frac{1}{\sqrt{5}}$, assumed at $(\frac{\pi}{2},\pi)$. Since there are effectively two sweeps for each θ, per W-cycle, the convergence factor per W-cycle should be $(\frac{1}{\sqrt{5}})^2 = 0.2$. This crude analysis just happens to work in this case.

What happens when this method is applied with "finite difference" boundary conditions? The asymptotic convergence factor per W-cycle is at least 10^4. What is the cause of this divergence? With the "finite difference" boundary conditions, (5.1) is no longer true. Is there an approximation to $L_{-1}^{M-1} = (U_{-1}^h I_{M-1}^M)^* L^{M-1} U_{-1}^h I_{M-1}^M$, $h = h_M$, that can be made in the finite difference case that will cure divergence? An answer is provided by examining numerically the operators L_{-1}^{M-1}, etc. in the variational approach. L_{-1}^{M-1} looks as if it were derived from imposing Dirichlet boundary conditions; the same is true for any L if U_{-1}^h, $h = h_j$, has been used in constructing it or if one of its predecessors was constructed using U_{-1}^h, $h = h_k$, $k > j$. This suggests that Dirichlet boundary conditions should be imposed in the appropriate places in Fig. 4. Dirichlet "finite difference" boundary conditions can be derived by using fictitious cells, writing down a difference approximation to $U = 0$ (e.g., $\frac{1}{2}(U_{1,j}+U_{2,j}) = 0$, $2 \leq j \leq 25$) and then eliminating the fictitious cells in terms of the interior cells. Two typical cases are: at (2,2), $-\Delta_h^{sk}U_{i,j} = \frac{1}{2h^2}(U_{i,j+1}+U_{i+1,j}-U_{i+1,j+1}+ 3U_{i,j})$ and at (2,j), $2 < j < 25$, $-\Delta_h^{sk}U_{i,j} = \frac{1}{2h^2}(U_{i,j-1}+ U_{i,j+1}-U_{i+1,j+1} - U_{i+1,j-1}+4U_{i,j})$. We use the notation L_1^k and L_{-1}^k to denote $-\Delta_{h_k}^{sk} + 10^{-4}$ with zero Neumann and zero Dirichlet "finite difference" boundary conditions, respectively, and we have Fig. 5. The asymptotic convergence factor per w-cycle is 0.21.

One remaining problem is that the computation of I^* is quite expensive; for each coarse grid cell center, I^* involves the weighted sum of sixteen residuals. In [4] what was used instead was

$$R_{k,\ell}^c = \tfrac{1}{4}(r_{i,j} + r_{i+1,j} + r_{i,j+1} + r_{i+1,j+1}) \quad , \quad (5.3)$$

where the (k,ℓ)th cell center on the coarse grid is in the center of the (i,j), (i+1,j), (i,j+1), (i+1,j+1) fine cell center. We use (5.3) here for $J_{3,1}^2$ and $J_{2,1}^1$, the replacements for $(I_{2,1}^3)^*$ and $(I_{1,1}^2)^*$. $J_{2,-1}^1$ and $J_{3,-1}^2$, the replacements for $I_{1,-1}^2$ and $I_{2,-1}^3$ are given by

$$R_{k,\ell}^c = \tfrac{1}{4}(-r_{i,j} + r_{i+1,j} + r_{i,j+1} - r_{i+1,j+1}) \quad .$$

The asymptotic convergence factor per W-cycle with these changes instituted is 0.18.

Can we dispense with the smoothing operator S^{k-1} in this method? The result of doing so is an asymptotic convergence factor greater than 10^6. Since (5.2) thus appears to be so important, can we just use it coupled with the usual multigrid method (with no "crazy" interpolations and operators)? The result of doing so is an asymptotic convergence factor (per V-cycle) of 0.93.

6. THE RAY MULTIGRID METHOD APPLIED TO THE PRESSURE ITERATION IN SALE.

A good question is why one should be interested in trying to accelerate the pressure iteration in SALE. If it has the bad feature of annihilating the (π,π) frequency, why not abandon it for something better? (An attempt was made to do exactly that in [4], but the resulting method was not better.) From a finite element point of view the method in SALE results from using piecewise bilinear elements and midpoint quadrature. If the method were implemented as a finite element method and four Gauss point quadrature were used, then the (π,π) frequency would not be annihilated; however, the calculation would be four times as expensive. In fact, one advocate of finite elements in fluid calculations confessed to me that he did not believe that his code could compete with the Lagrangian codes unless it used midpoint quadrature. On a given grid, the (π,π) frequency is badly approximated anyway, and while four Gauss point quadrature will give asymptotically more accurate answers, the goal of many fluid calculations is a qualitative representation of the flow; for such calculations, using midpoint quadrature may give accurate enough answers at less than one fourth the cost ("less than" since with midpoint quadrature the arithmetic simplifies considerably, a fact that SALE exploits to save storage by never explicitly forming a matrix). The "hourglass" instability can be controlled with smoothing of the (π,π) frequency - referred to as an "alternate node coupler" in SALE. One must exercise care not to introduce too much smoothing, and aside from philosophical qualms concerned with ad hoc smoothing, there is no reason to fault this procedure.

The paper [4] began with the application of multigrid to SOLA, an incompressible Eulerian code. For the problem under consideration, we used the residual weighting (5.3), which gave rise to a convergence factor per work unit of 0.64 as opposed to the supposed possible one of 0.595. Brandt devised another weighting scheme which used (5.3) in the interior and another weighting near the boundary. "Brandt's" weighting led to a convergence factor of 0.595 per work unit and was thus retained in the work for SOLA-ICE and SALE. However, in computational studies for this paper, we discovered that "Brandt's" weighting gave much worse results than (5.3) for problems with discontinuous right-hand sides. On the average, (5.3) appears to be better and is the weighting which we now recommend. We accept the blame for insufficient testing.

One problem for which (5.3) is much better than "Brandt's" weighting is the Rayleigh-Taylor problem worked with SALE in [4]. As commented in Sec. 1, we changed the differencing in the pressure iteration in SALE in [4]. Here we report on multigrid applied to SALE using the original differencing in SALE and the residual weighting (5.3). We take $M = 3$, $G^3 = 12$ cells $\times$ 12 cells grid, $IM = 2$, $IU = 2$, and $ID = 2$; this is in contrast to [4] where the accomodative mode of multigrid was used. The other parameters are the same except for the sound speed squared which is 2,000 in this paper and 20,000 in [4]. On the coarsest grid we use the iteration of the code coupled with the "constant addition" iteration; the same is true for the single grid calculation; see [4] for details. As in [4] we facilitate comparison between single grid and multigrid by using the following time steps:

$$\Delta t = \begin{cases} .5, & t < 15 \\ .25, & 15 \le t < 19 \\ .125, & 19 \le t < 23 \\ .0625, & 23 \le t < 25 \end{cases} .$$

No rezoning is used, and the grid distorts until a "bowtie" forms after t = 25. and the computation can no longer proceed; Fig. 6 shows a mesh near the end of the calculation.

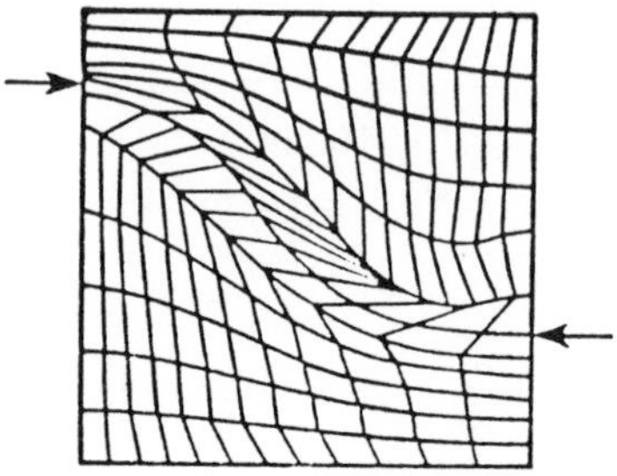

Fig. 6. Mesh used in Rayleigh-Taylor calculation at t=24.0.

(We remark that no "alternative node coupler" is used in this calculation. We tried using it, and it made little difference, as least for this problem. Also, we used an overrelaxation factor $\omega = 1.5$ for the single grid calculation as in [4].) Table 1 summarizes the comparison.

This is not exactly a smashing success for the multigrid method. Can the ray multigrid method be applied? One stumbling block is boundary conditions. The boundary conditions in SALE are applied by specifying u and v, the horizontal and vertical velocities. For the above problem u is specified on the top and bottom boundaries; this is equivalent to specifying $\frac{\partial p}{\partial \nu}$ on all boundaries. As remarked before, on a rectangular grid, the approximation to the Laplacian with $\frac{\partial p}{\partial \nu}$ specified on the boundary is $\Delta_h^{sk}p$ with the "finite difference" approximation to the boundary conditions. It can be checked that specifying v = 0 on the left and right boundaries and u = 0 on the bottom and top boundaries is equivalent to specifying p = 0 on all boundaries and that the approximation to the Laplacian with p = 0 specified on the boundary is $\Delta_h^{sk}p$ with the "finite difference" approximation to p = 0 on the boundary. Thus the algorithm displayed in Fig. 5 is implemented by specifying u[v] on the left and right boundaries and v[w] on the bottom and top boundaries for L_i^k if i = 1[-1].

Table 1. Comparison of single grid and regular multigrid in SALE.

Time	Total time spent iterating (MG/SG)	Total calculational time (MG/SG)	Fraction of calculation spent iterating (SG)	Fraction of calculation spent iterating (MG)	Convergence factor per work unit on last relaxation sweep, this time step (SG)	Convergence factor per work unit* on last cycle, this time step (MG)
t = 3.0	.96	.76	.63	.63	.82	.77
t = 15.0	.76	.77	.69	.66	.82	.67
t = 19.0	.69	.73	.76	.74	.85	.80
t = 23.0	.76	.79	.77	.76	.69	.80
t = 24.0	.83	.84	.79	.79	.84	.86
t = 25.0	.88	.89	.68	.69	.82	.89

* Only relaxation work is counted.

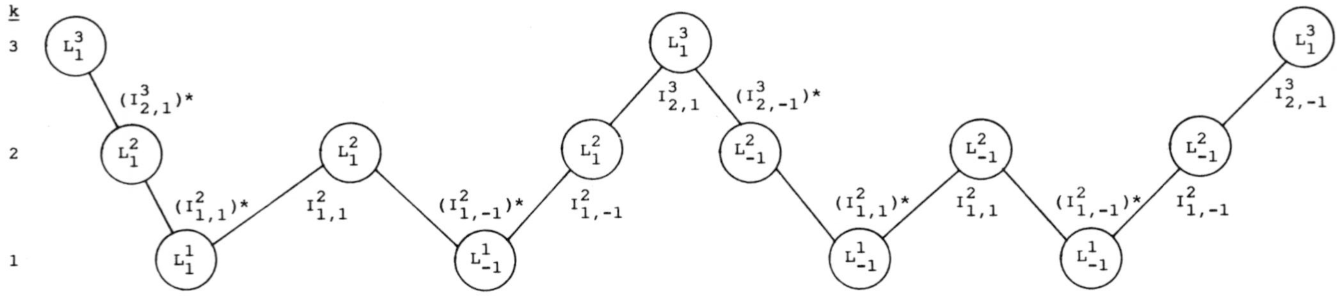

Fig. 5. W-cycle for ray multigrid with operator approximation and "finite difference" boundary conditions.

The constant addition algorithm is used on grid 1 only for L_1^1. We take IM = 2, IU = 2, ID = 2. The results, in Table 2, are promising for such a small problem. The convergence factor for the single grid calculation can be expected to increase with the number of unknowns, whereas the multigrid convergence factor should remain bounded. We avoided rezoning in this calculation to exhibit that multigrid can still function well on highly distorted grids.

Although multigrid has been successfully applied in this situation without cheating, that is, without changing the differencing in SALE, the resulting algorithm is far from robust. If the mesh has $\Delta x \ll \Delta y$ or $\Delta y \ll \Delta x$ in part or all of the mesh, then the appropriate line relaxation is necessary to achieve good multigrid convergence. Moreover, if the density jumps by orders of magnitude across internal interfaces, then the above algorithm will fail.[2] Both of these situations appear to require the matrix explicitly so that one of the methods in section 4 may be used; any of these methods will require a restructuring of the iteration used in SALE. For example, one possibility is to form a nine point numerical Jacobian and to use one of the methods in section 4 to solve the resulting linear system. We have begun investigating this approach and have encountered pitfalls. The first is that in the above problem the numerical (nine point) Jacobian is not positive definite. The second is that even if the problem is changed to have constant density, the numerical Jacobian is so nearly singular that problems arise in using Newton's method or damped Newton (where the Newton step is excruciatingly small). Experts have assured us that both these difficulties can be overcome, and if such is the case, perhaps the results will appear in "Son of Multigrid Semi-Implicit Hydrodynamics."

Table 2. Comparison of single grid and ray multgrid in SALE.

Time	Total time spent iterating (MG/SG)	Total calculational time (MG/SG)	Fraction of calculation spent iterating (MG)	Convergence factor per W-cycle and work unit* on last cycle, this time step (MG)
t = 3.0	.64	.60	.55	.05, .64
t = 15.0	.61	.67	.58	.06, .67
t = 19.0	.60	.66	.62	.17, .77
t = 23.0	.66	.71	.71	.19, .79
t = 24.0	.68	.72	.73	.10, .72
t = 25.0	.68	.72	.74	.14, .75

* Only relaxation work is counted.

ACKNOWLEDGMENT. As in [6], we make the canonical acknowledgment.

REFERENCES

1. Alcouffe, R. E., The multigrid method for solving the two-dimensional multigroup diffusion equation, Proc. of Advances in Reactor Computations, March 28-30, 1983, Salt Lake City, UT, 340-351.

2. ______________, A. Brandt, J. E. Dendy, Jr., and J. W. Painter, The multigrid method for the diffusion equation with strongly discontinuous coefficients, SIAM J. Sci. Stat. Comput. 2, (1981), 430-454.

3. Brandt, A., Multi-level adaptive solutions to boundary value problems, Math. Comp. 31 (1977), 333-390.

4. Brandt, A., J. E. Dendy, Jr., and H. Ruppel, The multi-grid method for semi-implicit hydrodynamics codes, J. Comp. Phys. 34 (1980), 348-370.

5. Dendy, J. E., Jr., Black box multigrid, J. Comp. Phys. 48 (1982), 366-386.

6. ______________, Black box multigrid for nonsymmetric problems, (to appear).

7. Donea, J., S. Guiliani, K. Morgan, and L. Quartapelle, The significance of chequerboarding in a Galerkin finite element solution of the Navier-Stokes equation, Int. J. Numer. Meth. in Eng'g. 17 (1981), 790-795.

8. Ta'asan, S., Multi-grid methods for highly oscillatory problems, research report, Weizman Institute, Rehovot, Israel, 1981.

This work was performed under the auspices of the U.S. Department of Energy.

Theoretical Division, MS-B284
Los Alamos National Laboratory
Los Alamos, NM 87545

NUMERICAL SOLUTION OF LARGE NONLINEAR BOUNDARY VALUE PROBLEMS BY QUADRATIC MINIMIZATION TECHNIQUES

Roland Glowinski and Patrick LeTallec

1. INTRODUCTION

The objective of this paper is to describe the numerical treatment of large highly nonlinear two or three dimensional boundary value problems by quadratic minimization techniques. These techniques, mostly described in [1], were developed and applied to such situations by various research groups at INRIA (Institut National de Recherche en Informatique et Automatique), at AMD/BA (Avions Marcel Dassault et Breguet Aviation) or at L.C.P.C. (Laboratoire Central des Ponts et Chaussées).

In all the different situations where these techniques were applied, the methodology remains the same and is organized as follows :

(i) Derive a variational formulation of the original boundary value problem, and approximate it by Galerkin methods ;

(ii) Transform this variational formulation into a quadratic minimization problem (least squares methods) or into a sequence of quadratic minimization problems (augmented lagrangian decomposition) ;

(iii) Solve each quadratic minimization problem by a conjugate gradient method with preconditioning, the preconditioning matrix being sparse, positive definite, and fixed once for all in the iterative process.

This paper will illustrate the methodology above on two different examples : the description of least squares solution methods and their application to the solution of the unsteady Navier-Stokes equations for incompressible viscous fluids ; the description of augmented lagrangian decomposition techniques and their application to the solution of equilibrium problems in finite elasticity.

ISBN 0-12-546080-5

2. LEAST SQUARES SOLUTION OF A NONLINEAR MODEL PROBLEM

In order to introduce the techniques which lead to the solution of nonlinear boundary value problems by least squares and conjugate gradient methods, we shall first consider the solution of a simple nonlinear Dirichlet problem. In section 3, these methods will then be applied to the solution of the unsteady Navier-Stokes equations for incompressible viscous fluids.

2.1. Formulation of the model problem.

Let $\Omega \subset \mathbb{R}^N$ be a bounded domain with a smooth boundary $\Gamma = \partial\Omega$; let T be a nonlinear operator from $V = H_0^1(\Omega)$ to $V^\star = H^{-1}(\Omega)$, $V^\star$ being the topological dual space of V. Usual notations are used for the Sobolev spaces [2] ; in particular, $H_0^1(\Omega)$ denotes the space of real functions defined over Ω, square-integrable, with square integrable first order derivatives, and zero trace on Γ.

The nonlinear Dirichlet model problem is :

$$(2.1) \qquad \begin{cases} \underline{\text{Find}}\ u \in H_0^1(\Omega)\ \underline{\text{such that}} \\ -\Delta u - T(u) = 0\ \underline{\text{in}}\ H^{-1}(\Omega). \end{cases}$$

Observe that since u is in $H_0^1(\Omega)$, it has zero trace on Γ, therefore (2.1) is indeed a Dirichlet problem. We do not discuss here the existence and uniqueness properties of the solutions of (2.1) since we do not want to be very specific about operator T.

2.2. H^{-1} least squares formulation of the model problem.

Many least squares formulations of the model problem above can be proposed. Among them, a natural one, based on the norm which appears naturally in the model problem, consists in saying that the solutions of the model problem cancel the norm of $(\Delta u + Tu)$ in $V^\star$, and therefore minimize this norm over V. The model problem becomes :

$$(2.2) \qquad \underset{v \in V}{\text{Min}} \| \Delta v + Tv \|_{V^\star}$$

where the $V^\star$ norm is defined by duality

$$\| f \|_{V^\star} = \sup_{\substack{v \in V \\ \| v \| = 1}} \langle f, v \rangle ,$$

$\langle \cdot , \cdot \rangle$ denoting the duality pairing between $V^\star$ and V. But, since in our model problem, the Laplace operator Δ is an isometry between V and $V^\star$, we can reformulate (2.2) as :

$$(2.3) \qquad \underset{v \in V}{\text{Min}} \| \Delta^{-1}(\Delta v + Tv) \|_V^2 .$$

Taking into account the definition of the V norm $\left(V = H_0^1(\Omega)\right)$ and introducing the function $\xi(v)$ of V defined from v by

$$\Delta\xi = \Delta v + Tv \ \underline{\text{in}}\ H^{-1}(\Omega)$$

or equivalently by

$$\int_\Omega \underset{\sim}{\nabla}\xi\cdot\underset{\sim}{\nabla}w\,d\underset{\sim}{x} = \int_\Omega \underset{\sim}{\nabla}v\cdot\underset{\sim}{\nabla}w\,d\underset{\sim}{x} - \langle Tv,w\rangle,\ \forall w\in V,\ \xi\in V, \tag{2.4}$$

the model problem can finally be reformulated as

$$\boxed{\underset{v\in H_0^1(\Omega)}{\text{Min}} \left[\frac{1}{2}\int_\Omega |\underset{\sim}{\nabla}\xi(v)|^2 d\underset{\sim}{x}\right]} \tag{2.5}$$

where $\xi(v)$ is the solution of (2.4).

In (2.5), *we have obtained a formulation of our initial model problem as a quadratic minimization problem*. As soon as the initial problem (2.1) has a solution, this new formulation (2.5) is obviously equivalent to the initial one (2.1). By considering v as a *control vector*, ξ as a *state vector*, (2.4) as the *state equation* and the functional

$$J(v) = \frac{1}{2}\int_\Omega |\underset{\sim}{\nabla}\xi(v)|^2 d\underset{\sim}{x}$$

as a *cost function*, Problem (2.5) is transformed into the following optimal control problem :

<u>Find a control vector</u> u <u>which minimizes the cost function</u> J <u>over the set</u> V <u>of admissible control vectors</u>.

2.3. <u>Conjugate gradient solution of the least squares problem</u>. Problem (2.5) is a minimization problem. For its solution, we shall use a conjugate gradient algorithm. Due to its good performances, (POLAK [3], POWELL [4]), we have selected the *Polak-Ribière* version of the conjugate gradient algorithm for the solution of (2.5), which is

<u>Step 0</u> : <u>Initialisation</u>

$$u^0 \in H_0^1(\Omega)\ \textit{given}, \tag{2.6}$$

compute g^0 *in* $H_0^1(\Omega)$ *by*

$$-\Delta g^0 = J'(u^0)\ \textit{in}\ H^{-1}(\Omega) \tag{2.7}$$

and set

$$z^0 = g^0. \tag{2.8}$$

Then for $n \geq 0$, *assuming* u^n, g^n, z^n *known, compute* u^{n+1}, g^{n+1}, z^{n+1} *by*

Step 1 : Descent

(2.9) $$u^{n+1} = u^n - \lambda_n z^n,$$

where λ_n *is the solution of the one-dimensional minimization problem*

(2.10) $$\begin{cases} \lambda_n \in \mathbb{R} \\ J(u^n - \lambda_n z^n) \leq J(u^n - \lambda z^n), \quad \forall \lambda \in \mathbb{R}. \end{cases}$$

Step 2 : Construction of the new descent direction.

Define $g^{n+1} \in H_0^1(\Omega)$ *by*

(2.11) $$-\Delta g^{n+1} = J'(u^{n+1}) \ \underline{\text{in}}\ H^{-1}(\Omega).$$

and set

(2.12) $$\gamma_n = \left\{\int_\Omega \nabla g^{n+1} \cdot \nabla (g^{n+1} - g^n) dx\right\} \star \|g^n\|_V^{-2},$$

(2.13) $$z^{n+1} = g^{n+1} + \gamma_n z^n.$$

Go back to step 1 *with* $n = n+1$.

The two non-trivial steps of algorithm (2.6)-(2.13) are :

(i) The solution of the single variable minimization problem (2.10) ; the corresponding line search can be achieved by dichotomy or Fibonacci methods. Observe that each evaluation of $J(v)$ for a given argument v requires the solution of the linear Poisson problem (2.4) to obtain the corresponding ξ.

(ii) The calculation of g^{n+1} from u^{n+1} which requires the solution of two linear Poisson problems, namely (2.4) with $v = u^{n+1}$ and (2.11).

In order to check that (2.11) is indeed a linear Poisson problem, let us detail the calculation of g^{n+1}. By construction of J, its gradient is given by

$$\langle J'(v), w\rangle = \int_\Omega \nabla\xi(v) \cdot \nabla\eta(w) dx, \quad \forall w \in H_0^1(\Omega),$$

with η solution of

$$\Delta\eta(w) = \Delta w + T'(v) \cdot w \ \underline{\text{in}}\ H^{-1}(\Omega).$$

After elimination of η, this gives :

$$\langle J'(v), w\rangle = \int_\Omega \nabla\xi(v) \cdot \nabla w \, dx - \langle T'(v) \cdot w, \xi(v)\rangle, \quad \forall w \in H_0^1(\Omega).$$

Thus, problem (2.11) reduces to the following linear variational (Poisson) problem :

$$(2.14)\quad \begin{cases} \textit{Find } g^{n+1} \in H_0^1(\Omega) \textit{ such that, for any } w \textit{ in } H_0^1(\Omega), \textit{ we have :} \\ \int_\Omega \nabla g^{n+1} \cdot \nabla w \, d\underset{\sim}{x} = \int_\Omega \nabla \xi^{n+1}(u^{n+1}) \cdot \nabla w \, d\underset{\sim}{x} - \langle T'(u^{n+1})\ w,\ \xi^{n+1}(u^{n+1}) \rangle . \end{cases}$$

Remark 2.1. As stopping criterion for the conjugate gradient algorithm (2.6)-(2.13), we shall use

$$J(u^n) \leqslant \varepsilon \text{ or } \|g^n\|_{H_0^1(\Omega)} \leqslant \varepsilon,$$

where ε is a reasonably small positive number. □

Remark 2.2. It is clear from the above observations that an *efficient Poisson solver is the basic tool* for solving the model problem (2.1) by our conjugate gradient algorithm. Any numerical limitation (in size) for this algorithm will come from a limitation on the Poisson solver. □

Remark 2.3. The above methodology extends easily to the solution of many other nonlinear boundary value problems : arc length continuation methods ([5]), Von Karman equations for thin clamped plates ([6]), transsonic flow problems ([7]). The choice of (2.1) as a model problem was only made for clarity reasons. In the next section, we shall apply this methodology to the solution of the nonlinear elliptic system

$$\alpha \underset{\sim}{u} - \Delta \underset{\sim}{u} + (\underset{\sim}{u} \cdot \underset{\sim}{\nabla}) \underset{\sim}{u} = \underset{\sim}{f} \text{ on } \Omega$$

with $\underset{\sim}{u}$ a vector function defined on Ω with values in $\mathbb{R}^N$. Such a system is closely related to the solution of the time dependent Navier-Stokes equations by alternating direction methods. □

3. APPLICATION TO THE SOLUTION OF THE NAVIER-STOKES EQUATIONS

3.1. Formulation of the time dependent Navier-Stokes equations for incompressible viscous fluids. Let us consider a Newtonian incompressible viscous fluid. If Ω and Γ denote the region of the flow ($\Omega \subset \mathbb{R}^N$, $N = 2$ or 3) and its boundary, then this flow is governed by the Navier-Stokes equations which relate velocity and pressure inside the fluid to the external loads and which are :

$$(3.1)\quad \begin{cases} \dfrac{\partial u}{\partial t} - \nu \Delta \underset{\sim}{u} + (\underset{\sim}{u} \cdot \underset{\sim}{\nabla}) \underset{\sim}{u} + \underset{\sim}{\nabla} p = \underset{\sim}{f} \text{ in } \Omega, \\ \operatorname{div} \underset{\sim}{u} = 0 \text{ in } \Omega. \end{cases}$$

Above $\mathbf{u} : \Omega \rightarrow \mathbb{R}^N$ denotes the flow velocity, p the hydrostatic pressure, ν the viscosity of the fluid, $\mathbf{f}$ the density of external forces. Moreover, the notation $(\mathbf{u}\cdot\nabla)\mathbf{u}$ is a symbolic notation for the unsymmetric quadratic vector term

$$(\mathbf{u}\cdot\nabla)\mathbf{u} = \sum_{j=1}^{N} \left\{u_j \frac{\partial u_i}{\partial x_j}\right\}_{i=1,N}$$

which corresponds to the convection part of equation (3.1).

To fully characterize the flow, initial and boundary conditions must be imposed on $\mathbf{u}$. In the case of the airfoil A of Figure 3.1, we have, typically,

(3.2) $\mathbf{u}(\mathbf{x},t) = \mathbf{0}$ on ∂A (adhérence condition on the airfoil),

(3.3) $\mathbf{u}(\mathbf{x},t) = \mathbf{u}_\infty(t)$ at infinity,

(3.4) $\mathbf{u}(\mathbf{x},0) = \mathbf{u}_0(\mathbf{x})$ (initial condition).

Fig. 3.1

For a flow in a bounded region of $\mathbb{R}^N$, we may replace the boundary conditions (3.2) and (3.3) by

(3.5) $$\mathbf{u}(\mathbf{x},t) = \mathbf{g}(\mathbf{x}) \text{ on } \Gamma,$$

where, due to the incompressibility condition $\operatorname{div}\mathbf{u} = 0$, the given function $\mathbf{g}$ must satisfy

$$\int_\Gamma \mathbf{g}\cdot\mathbf{n}\, d\Gamma = 0 \quad (\mathbf{n} = \text{unit normal to } \Gamma).$$

On the above equations, we observe two main difficulties :

(i) the nonlinear term $(\mathbf{u}\cdot\nabla)\mathbf{u}$ in (3.1) ;

(ii) the incompressibility constraint $\operatorname{div}\mathbf{u} = 0$.

Using convenient alternating direction methods for the time discretization of the Navier-Stokes equations, we are able to uncouple these difficulties. Problem (3.1) with initial and boundary conditions (3.2)-(3.4) (respectively (3.4)-(3.5) for bounded domains) then reduces to a sequence of

(a) incompressible linear problems,

(b) compressible nonlinear problems to be solved by the least squares methods of Sec 2, that is via the solution of a sequence of linear strongly elliptic problems.

All the resulting linear problems will be associated to fixed matrices. Adequate algorithms can then be deviced for their numerical solutions, even for large size problems. At this stage domain decomposition techniques become very attractive.

For simplicity, we suppose from now on that Ω is bounded.

3.2. Time discretization by alternating direction methods. Let Δt be a time discretization step. The alternating direction method that was found to be computationally the most convenient for the time discretization of the Navier-Stokes equations, consists in replacing (3.1) by the following sequence of problems :

(3.6) *Let* $\mathbf{u}^0 = \mathbf{u}_0$;

then for $n \geqslant 0$ *and starting from* $\mathbf{u}^n$, *we solve successively*

$$(3.7)\quad \begin{cases} \dfrac{(\mathbf{u}^{n+1/4} - \mathbf{u}^n)}{\Delta t/4} \quad \dfrac{2\nu}{3}\Delta\mathbf{u}^{n+1/4} + \nabla p^{n+1/4} = \mathbf{f}^{n+1/4} + \dfrac{\nu}{3}\Delta\mathbf{u}^n - (\mathbf{u}^n\cdot\nabla)\mathbf{u}^n \ \underline{\text{in}}\ \Omega; \\ \text{div}\ \mathbf{u}^{n+1/4} = 0 \ \underline{\text{in}}\ \Omega, \\ \mathbf{u}^{n+1/4} = \mathbf{g}^{n+1/4} \ \underline{\text{on}}\ \Gamma; \end{cases}$$

$$(3.8)\quad \begin{cases} \dfrac{(\mathbf{u}^{n+3/4} - \mathbf{u}^{n+1/4})}{\Delta t/2} - \dfrac{\nu}{3}\Delta\mathbf{u}^{n+3/4} + (\mathbf{u}^{n+3/4}\cdot\nabla)\mathbf{u}^{n+3/4} = \mathbf{f}^{n+3/4} - \nabla p^{n+1/4} \\ \qquad\qquad + \dfrac{2\nu}{3}\Delta\mathbf{u}^{n+1/4} \ \underline{\text{in}}\ \Omega, \\ \mathbf{u}^{n+3/4} = \mathbf{g}^{n+3/4} \ \underline{\text{on}}\ \Gamma; \end{cases}$$

$$(3.9)\quad \begin{cases} \dfrac{\mathbf{u}^{n+1} - \mathbf{u}^{n+3/4}}{\Delta t/4} - \dfrac{2\nu}{3}\Delta\mathbf{u}^{n+1} + \nabla p^{n+1} = \mathbf{f}^{n+1} + \dfrac{\nu}{3}\Delta\mathbf{u}^{n+3/4} - (\mathbf{u}^{n+3/4}\cdot\nabla)\mathbf{u}^{n+3/4}, \\ \text{div}\ \mathbf{u}^{n+1} = 0 \ \underline{\text{in}}\ \Omega, \\ \mathbf{u}^{n+1} = \mathbf{g}^{n+1} \ \underline{\text{on}}\ \Gamma. \end{cases}$$

The notations $\underline{f}^j(\underline{x})$ and $\underline{g}^j(\underline{x})$ denote $\underline{f}(\underline{x},j\Delta t)$ and $\underline{g}(\underline{x},j\Delta t)$, and $\underline{u}^j(\underline{x})$ is an approximation for $\underline{u}(\underline{x},j\Delta t)$.

Remark 3.1 : Due to the symmetrization process that it involves, scheme (3.7)-(3.9) has a truncation error of $O(|\Delta t|^2)$. Although the linear steps (3.7) and (3.9) are identical, suppressing one of them would not be advisable : it would increase the truncation error with no real gain on the computation time, this computation time being mainly devoted to the nonlinear step (3.8).   □

Remark 3.2 : The decomposition of the operator $\Delta\underline{u}$ between the right and left sides of equations (3.7), (3.8) and (3.9) was done in order to involve the same linear operators in each step, which results in quite substantial computer core memory savings.   □

Remark 3.3 : We have introduced the alternating direction decomposition of (3.1) on the continuous problems because the formalism is simpler. But, of course, the same decomposition applies to any Galerkin finite dimensional variational formulation of the Navier-Stokes equations obtained, let's say, by finite element methods.   □

3.3. Least-squares-conjugate gradient solution of the nonlinear subproblems (3.8). At each full step of the alternating direction methods (3.7)-(3.9), we have to solve a nonlinear elliptic system of the following type :

$$(3.10) \qquad \begin{cases} \alpha\underline{u} - \nu\,\Delta\underline{u} + (\underline{u}\cdot\underline{\nabla})\underline{u} = \underline{f} \ \underline{\text{in}}\ \Omega, \\ \underline{u} = \underline{g} \ \underline{\text{on}}\ \Gamma. \end{cases}$$

Once the Laplace operator of Sec. 2 has been replaced by the operator $\alpha\mathrm{Id} - \nu\Delta$, the application to (3.10) of the least squares methodology of Sec. 2 transforms (3.10) into the following minimization problem :

$$(3.11) \qquad \boxed{\ \underset{\underline{u}\in V_g}{\mathrm{Min}}\ J(\underline{u}) = \frac{1}{2}\int_\Omega \{\alpha\,|\underline{y}(\underline{u})|^2 + \nu\,|\underline{\nabla}\underline{y}(\underline{u})|^2\}d\underline{x},\ }$$

where the vector field $\underline{y}(\underline{u})$ in V_0 is the solution of

$$(3.12) \qquad \boxed{\ \alpha\underline{y} - \nu\,\Delta\underline{y} = \alpha\underline{u} - \nu\,\Delta\underline{u} + (\underline{u}\cdot\underline{\nabla})\underline{u} - \underline{f}\ } \qquad \text{in } V_0^\star,$$

the space V_0 and the set V_g being defined respectively as

$$V_0 = \left(H_0^1(\Omega)\right)^N, \quad V_g = \left\{\underline{u}\in\left(H^1(\Omega)\right)^N,\ \underline{u} = \underline{g}\ \text{on}\ \Gamma\right\}.$$

Thus, the solution of Step (3.8) in the alternating direction time discretization of the Navier-Stokes equations reduces to the solution of the minimization problem (3.11) by conjugate gradient methods. Applied to (3.11), the Polak-Ribière version of the conjugate gradient algorithm is again given by the succession of steps (2.6) to (2.13), the Laplace operator being replaced by $\alpha Id - \nu\Delta$. In particular, step (2.11) which calculates the gradient direction in V_0 of the functional J at point $\underset{\sim}{u}^{n+1}$ is now given as the following *linear variational problem* :

$$(3.13)\quad \begin{cases} \underline{\text{Find } \underset{\sim}{g}^{n+1} \in V_0, \text{ such that, for any } \underset{\sim}{w} \text{ in } V_0, \text{ we have}} : \\ \int_\Omega \{\alpha \underset{\sim}{g}^{n+1}\cdot\underset{\sim}{w} + \nu \underset{\sim}{\nabla}\underset{\sim}{g}^{n+1}\cdot\underset{\sim}{\nabla}\underset{\sim}{w}\}d\underset{\sim}{x} = \int_\Omega \{\alpha \underset{\sim}{y}(\underset{\sim}{u}^{n+1})\cdot\underset{\sim}{w} + \nu\underset{\sim}{\nabla}\underset{\sim}{y}(\underset{\sim}{u}^{n+1})\cdot\underset{\sim}{\nabla}\underset{\sim}{w}\}d\underset{\sim}{x} \\ \qquad + \int_\Omega \{\underset{\sim}{y}(\underset{\sim}{u}^{n+1})\cdot(\underset{\sim}{w}\cdot\underset{\sim}{\nabla})\underset{\sim}{u}^{n+1} + \underset{\sim}{y}(\underset{\sim}{u}^{n+1})\cdot(\underset{\sim}{u}^{n+1}\cdot\underset{\sim}{\nabla})\underset{\sim}{w}\}d\underset{\sim}{x}. \end{cases}$$

Each iteration of the conjugate gradient algorithm applied to the solution of (3.11) finally requires the solution of four linear systems associated to the operator $\alpha Id - \nu\Delta$, that is

(i) one for computing $\underset{\sim}{y}(\underset{\sim}{u}^{n+1})$ through (3.12),

(ii) one for computing $\underset{\sim}{g}^{n+1}$ though (3.13),

(iii) two to obtain the coefficients of the quartic polynomial

$$\lambda \to J(\underset{\sim}{u}^n - \lambda\underset{\sim}{w}^n).$$

In practive, the solution of the one-dimensional line search problem can be done very efficiently since it is equivalent to finding the roots of a single variable cubic polynomial whose coefficients are known. The solution of each system associated to $\alpha Id - \nu\Delta$ corresponds to the solution of N independent scalar Dirichlet problems associated to the same operator. And the conjugate gradient algorithm appears to be very efficient in this situation : usually three iterations suffice to reduce the cost function by a factor of 10^4 to 10^6. Therefore, the whole solution of Step (3.8) by the techniques indicated in this paragraph is not costly, nor for its implementation, neither for its computational run time.

3.4. Solution of the quasi-Stokes problems (3.7) and (3.9). These linear equations, which appear at each full step of the alternating direction method (3.7)-(3.9), involve two unknowns (velocity and pressure) and are of the type :

$$(3.14)\quad \begin{cases} \alpha\underset{\sim}{u} - \nu\,\Delta\underset{\sim}{u} + \underset{\sim}{\nabla}p = \underset{\sim}{f} \ \underline{\text{in}}\ \Omega, \\ \operatorname{div}\underset{\sim}{u} = 0 \text{ in } \Omega,\ \underset{\sim}{u} = \underset{\sim}{g}\ \underline{\text{on}}\ \Gamma; \end{cases}$$

Many existing solvers can be used for this problem. Nevertheless, direct methods of Taylor-Hood type [8] are not very adequate here, because they are expensive in computer memory core : they imply a direct inversion of the matrix obtained via a Galerkin variational approximation of (3.14) ; this matrix is not positive and is very large (both velocities and pressure are present), therefore difficult to invert in large scale computations

For (3.14), one can also think of other methods such as Augmented Lagrangian techniques (see Sec. 4 and 5) or boundary methods and we refer to [1 chap. 7] for more details. But, in fact, least squares and conjugate gradient methods can also be used here. Indeed, (3.14) can be reformulated as the following minimization problem :

$$(3.15) \qquad \operatorname*{Min}_{p \in L^2(\Omega)/\mathbb{R}} \int_\Omega \frac{1}{2}[\operatorname{div}(\underset{\sim}{u}(p))]^2 d\underset{\sim}{x}$$

where $\underset{\sim}{u}(p)$ in V_g is the solution of

$$(3.16) \qquad \alpha\underset{\sim}{u} - \Delta\underset{\sim}{u} = \underset{\sim}{f} - \underset{\sim}{\nabla}p \text{ in } V_0^\star.$$

For the solution of (3.15), we simply apply the Polak-Ribière conjugate gradient algorithm described before. Observe then that the linear operator acting on $\underset{\sim}{u}$ is the same operator $(\alpha \mathrm{Id} - \nu\Delta)$ which appears in the nonlinear problem (3.8), which is very convenient for the practical implementation of this algorithm.

Remark 3.4. For very large problems, when the finite element approximation of the Navier-Stokes equations involve several ten thousands of unknowns, it might be useful to split first Problems (3.7) or (3.9) into smaller scale problems of the same type, obtained by domain decomposition techniques (see [9] for more details). □

3.5. Numerical experiments. We illustrate the numerical techniques discussed in the above sections by presenting the results of numerical experiments where these techniques have been used to simulate several incompressible viscous flows modeled by the Navier-Stokes equations. The experiment presented here concerns an unsteady flow around and inside a nozzle at high incidence and at Reynolds number 750 (the characteristic length being the distance between the nozzle walls). Figures 3.2 to 3.5 represent the streamlines at $t = 0$, $t = .2$, $t = .4$, $t = .6$ respectively, showing clearly the creation and the motion of eddies, inside and behind the nozzle.

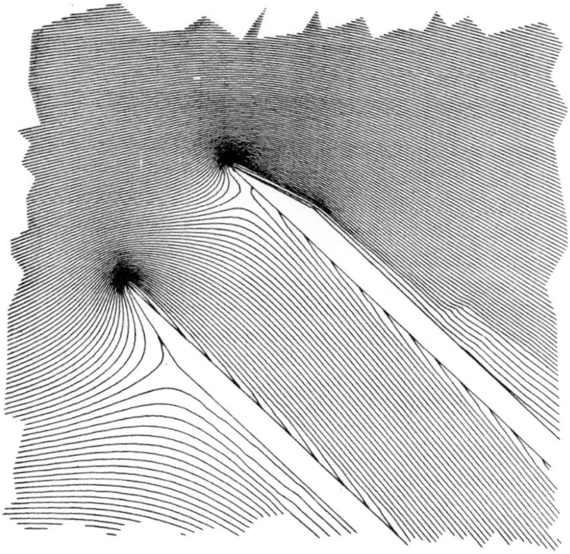

Figure 3.2

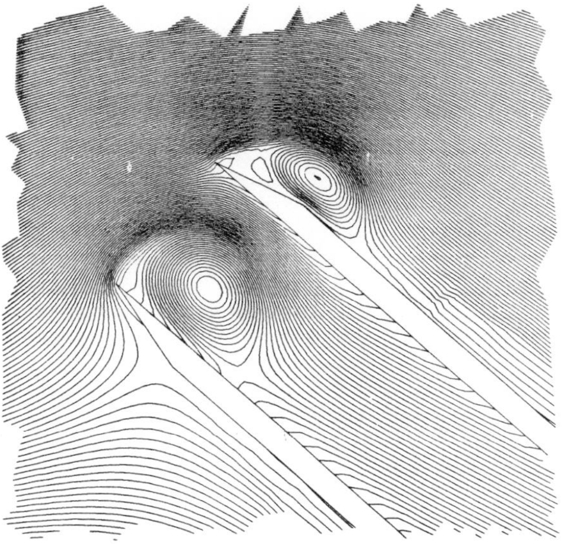

Figure 3.3

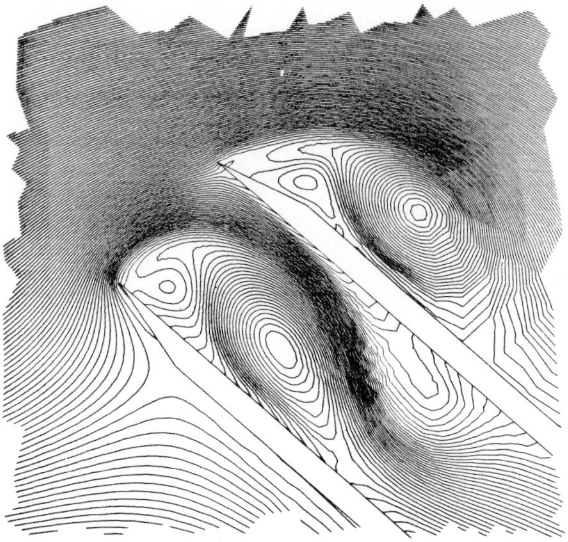

Figure 3.4

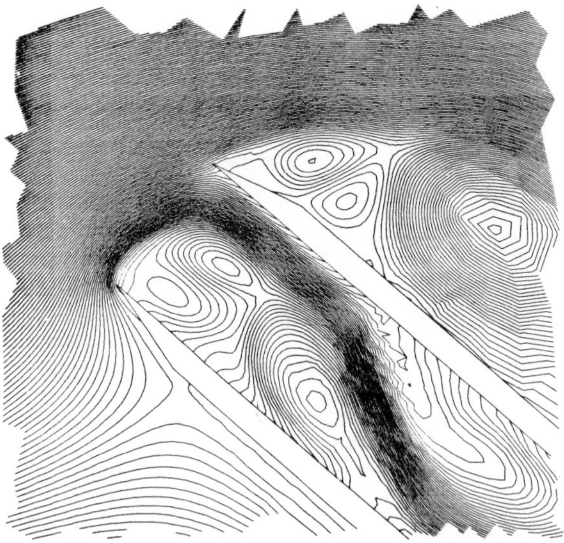

Figure 3.5

4. DECOMPOSITION METHODS BY AUGMENTED LAGRANGIANS

The main goal of this section is to give a brief account of solution methods for variational problems when some decomposition property holds ; introducing a convenient augmented lagrangian, we obtain solution methods taking full advantage of the special structure of the problem under consideration. We shall first consider in this section the solution by augmented lagrangians of a simple model problem, before considering in Sec. 5 the application of these techniques in three dimensional Finite Elasticity.

4.1. Formulations of the model problem. Let $\Omega \subset \mathbb{R}^2$ be a bounded domain with a smooth boundary $\Gamma = \partial\Omega$, and consider the following model problems (with $1 < p < +\infty$) :

$$\text{(4.1)} \qquad \begin{cases} -\operatorname{div}(|\nabla u|^{p-2}\nabla u) = f \ \underline{\text{in}}\ \Omega, \\ u = 0 \ \underline{\text{on}}\ \Gamma_1, \\ |\nabla u|^{p-2}\nabla u \cdot n = g \ \underline{\text{on}}\ \Gamma_2,\ \Gamma_1 \cap \Gamma_2 = \phi,\ \Gamma_1 \cup \Gamma_2 = \Gamma. \end{cases}$$

Such problems, discussed in [10], appear for example in the study of Norton viscoplastic fluids flowing viscously in a cylindrical duct. They can be written under variational form as

$$\text{(4.2)} \qquad \underset{v \in V}{\operatorname{Min}}\ J(v) = \int_\Omega \frac{1}{p}|\nabla v|^p dx - \int_\Omega f\, v\, dx - \int_{\Gamma_2} g\, v\, ds,$$

with

$$\text{(4.3)} \qquad V = \{v \in W^{1,p}(\Omega),\ v = 0 \ \underline{\text{on}}\ \Gamma_1\}.$$

Observe now that the functional $J(v)$ can be naturally decomposed into

$$\text{(4.4)} \qquad \begin{cases} J(v) = \mathcal{F}(Bv) + \mathcal{G}(v) \ \underline{\text{with}} \\ Bv = \nabla v,\ \mathcal{F}(G) = \int_\Omega \frac{1}{p}|G|^p dx,\ \mathcal{G}(v) = -\int_\Omega f\, v\, dx - \int_{\Gamma_2} g\, v\, ds. \end{cases}$$

Therefore, Problem (4.1) can also be formulated as

$$\text{(4.5)} \qquad \begin{cases} \underline{\text{Find}}\ \{u, F\} \in W\ \underline{\text{such that}} \\ j(u,F) \leq j(v,G),\ \forall \{v,G\} \in W \end{cases}$$

where the space W and the functional j are defined by

$$\text{(4.6)} \qquad W = \{\{v,G\} \in V \times H,\ G = Bv\},\ H = (L^p(\Omega))^N,$$

$$\text{(4.7)} \qquad j(v,G) = \mathcal{F}(G) + \mathcal{G}(v).$$

Problems (4.2) and (4.5) are indeed equivalent but (4.5) has, in some sense a simpler structure than (4.2), despite the fact that it contains an extra variable. This is because the linear relation $Bv - \underset{\sim}{G} = 0$ can be efficiently treated using simultaneously penalty and lagrange multipliers methods of solutions, this via an appropriate augmented lagrangian [11].

4.2. An augmented lagrangian associated to (4.5). Let R be a strictly positive real constant, and define the augmented lagrangian $\mathcal{L}$ from $V\times H\times H^{\star}$ ($H^{\star}$ topological dual space of H) into $\mathbb{R}$ by

$$(4.8)\quad \begin{cases} \mathcal{L}(v,\underset{\sim}{G},\underset{\sim}{\mu}) = \mathcal{F}(\underset{\sim}{G}) + \mathcal{G}(v) + \frac{R}{2}\|Bv-\underset{\sim}{G}\|^2_{0,2} - \langle\underset{\sim}{\mu},Bv-\underset{\sim}{G}\rangle \\ \quad = \int_\Omega \frac{1}{p}|\underset{\sim}{G}|^p d\underset{\sim}{x} - \int_\Omega f\,v\,d\underset{\sim}{x} - \int_{\Gamma_2} g\,v\,ds + \int_\Omega \{\frac{R}{2}|\underset{\sim}{\nabla}v-\underset{\sim}{G}|^2 + \underset{\sim}{\mu}\cdot(\underset{\sim}{\nabla}v-\underset{\sim}{G})\}d\underset{\sim}{x}. \end{cases}$$

Consider now the problem

(4.9) Find $\{\{u,\underset{\sim}{F}\},\underset{\sim}{\lambda}\}$ saddle point of $\mathcal{L}$ over the space $\{V\times H\}\times H^{\star}$

or, in other words,

Find $\{\{u,\underset{\sim}{F}\},\underset{\sim}{\lambda}\} \in \{V\times H\}\times H^{\star}$ such that

$$(4.10)\quad \mathcal{L}(u,\underset{\sim}{F},\underset{\sim}{\lambda}) \leq \mathcal{L}(v,\underset{\sim}{G},\underset{\sim}{\lambda}),\ \forall\{v,\underset{\sim}{G}\}\in V\times H,$$

$$(4.11)\quad \mathcal{L}(u,\underset{\sim}{F},\underset{\sim}{\lambda}) \geq \mathcal{L}(u,\underset{\sim}{F},\underset{\sim}{\mu}),\ \forall\underset{\sim}{\mu}\in H^{\star}.$$

Problems (4.9) and (4.5) are *equivalent*. Indeed, let $\{\{u,\underset{\sim}{F}\},\underset{\sim}{\lambda}\}$ be a solution of (4.9). From (4.11), necessarily we have $\underset{\sim}{F} = Bu$, which means that $\{u,\underset{\sim}{F}\}\in W$. Then, if we write (4.10) with $\{v,\underset{\sim}{G}\}$ in W we obtain

$$j(u,\underset{\sim}{F}) \leq j(v,\underset{\sim}{G}),\ \forall\{v,\underset{\sim}{G}\}\in W,$$

which means precisely that $\{u,\underset{\sim}{F}\}$ is a solution of (4.5). Conversely, if $\{u,\underset{\sim}{F}\}$ is a solution of (4.5), by denoting $\underset{\sim}{\lambda} = \mathcal{F}'(\underset{\sim}{F})$, one checks easily that $\{\{u,\underset{\sim}{F}\},\underset{\sim}{\lambda}\}$ is a solution of (4.9). Observe that in this proof, the penalty term $\frac{R}{2}\|Bv-\underset{\sim}{G}\|^2_{0,2}$ plays no role. But its role is fundamental in accelerating the convergence of the numerical algorithms used for the solution of the saddle-point problem (4.9).

4.3. An Uzawa algorithm for solving (4.9). In Sec. 4.2, we have replaced our initial model problem (4.1) by the *equivalent saddle-point formulation* (4.9). A basic algorithm for the solution of this last problem combines an Uzawa algorithm for the solution of the saddle-point problem and a block relaxation algorithm for the solution of the minimization subproblem associated to the primal variable $\{v,\underset{\sim}{G}\}$. This leads to the following algorithm

(4.12) $$\{\underset{\sim}{\lambda}^0, u^{-1}\} \in H^{\star} \times V \textit{ is given ;}$$

then, for $n \geq 0$, u^{n-1} *and* $\underset{\sim}{\lambda}^n$ *being known, we compute* $\{u^n, \underset{\sim}{F}^n\}$ *by block relaxation,* i.e. *by*

(4.13) $$\textit{setting } u_0^n = u^{n-1},$$

and by computing sequentially u_k^n *and* $\underset{\sim}{F}_k^n$ *by solving*

(4.14) $$\mathcal{L}(u_{k-1}^n, \underset{\sim}{F}_k^n, \underset{\sim}{\lambda}^n) \leq \mathcal{L}(u_{k-1}^n, \underset{\sim}{G}, \underset{\sim}{\lambda}^n), \quad \forall \underset{\sim}{G} \in H,$$

(4.15) $$\mathcal{L}(u_k^n, \underset{\sim}{F}_k^n, \underset{\sim}{\lambda}^n) \leq \mathcal{L}(v, \underset{\sim}{F}_k^n, \underset{\sim}{\lambda}^n), \quad \forall v \in V \;;$$

Once $\{u^n, \underset{\sim}{F}^n\}$ *known, the lagrange multiplier* $\underset{\sim}{\lambda}^n$ *is updated by*

(4.16) $$\underset{\sim}{\lambda}^{n+1} = \underset{\sim}{\lambda}^n - \mathcal{R}_z\left(R(Bu^n - \underset{\sim}{F}^n)\right),$$

$\mathcal{R}_z$ *being the Riesz mapping sending* H *onto its dual* $H^{\star}$.

Remark 4.1. Many variants of the above algorithm exists ; they are described e.g. in [11 Chap. 3]. □

Remark 4.2. Under quite reasonable assumptions on $\mathcal{F}$, $\mathcal{G}$ or B, satisfied for example in the case of our model problems (4.2), the convergence of the above algorithm towards a solution $\{\{u, \underset{\sim}{F}\}, \underset{\sim}{\lambda}\}$ of (4.9) can be proved ([11 Chap. 3]) :

$$\begin{cases} \lim \{u^n, \underset{\sim}{F}^n\} = \{u, \underset{\sim}{F}\} & \underline{\text{strongly in}}\ V \times H, \\ \lim \underset{\sim}{\lambda}^n = \underset{\sim}{\lambda} & \underline{\text{weakly}}\ \text{in } H^{\star}. \end{cases}$$ □

Remark 4.3. It is interesting to further analyze the structure of the subproblems (4.14) and (4.15) which appear at each full step of the Uzawa algorithm used for the solution of our model problem (formulated as the saddle-point problem (4.9)). First, (4.14), whose unknown is $\underset{\sim}{F}$, involves no spatial derivatives of $\underset{\sim}{F}$; if an adequate discretization of H is used, it reduces to a family of independent algebraic pointwise problems of the type

(4.17) $$\underset{\underset{\sim}{G} \in \mathbb{R}^2}{\text{Min}} \left\{\frac{1}{p}|\underset{\sim}{G}|^p + \frac{R}{2}|\underset{\sim}{G} - Bu_{k-1}^n|^2 + \underset{\sim}{\lambda}^n \cdot \underset{\sim}{G}\right\}.$$

On the other hand, (4.15) is a global quadratic minimization problem, given by

(4.18) $$\underset{\underset{\sim}{v} \in V}{\text{Min}} \int_{\Omega} \frac{R}{2}|\underset{\sim}{\nabla} v - \underset{\sim}{F}_k^n|^2 d\underset{\sim}{x} - \int_{\Omega} \underset{\sim}{\lambda}^n \cdot \underset{\sim}{\nabla} v \, d\underset{\sim}{x} - \int_{\Omega} f\, v\, d\underset{\sim}{x} - \int_{\Gamma_1} g\, v\, ds.$$

Therefore, we have achieved a decomposition of our model problems into a sequence of pointwise algebraic problems which can be easily solved by Newton's type methods in $\mathbb{R}^2$ (in fact, they can be reduced to nonlinear problems in $\mathbb{R}$), and of global quadratic minimization problems, for which the conjugate gradient solution methods of Sec. 2 can be used. □

Remark 4.4. Although not critical, the choice of good values for the penalty coefficient R is complicated : theoretically the speed of convertence of algorithm (4.12)-(4.16) increases with R, but the conditioning of Problem (4.15) deteriorates as R increases. □

5. APPLICATION TO FINITE ELASTICITY

5.1. Generalities. The application of augmented lagrangian techniques to the solution of equilibrium problems in finite elasticity encounters three types of difficulties :

(i) The choice of reasonable constitutive laws ;

(ii) The choice of the right functional framework for the problem decomposition (there is no true convexity in hyperelasticity) ;

(iii) The derivation of adequate iterative methods for the pointwise solution of the algebraic problems appearing after decomposition.

Since it corresponds to our more recent numerical results, we will concentrate herein on the case of compressible hyperelastic bodies. The other case, concerning incompressible bodies, has already been extensively described in [12], [13] or [14] and is quite similar.

5.2. Formulations of equilibrium problems in compressible hyperelasticity. The considered problem consists in determining the final equilibrium position of an hyperelastic compressible body subjected to large deformations through the application of given external loads and imposed boundary displacements. We label any particle $\underset{\sim}{x}$ of the body by its position in a *stress free reference configuration* (lagrangian coordinates) and relate both $\underset{\sim}{x}$ and the displacement $\underset{\sim}{u}(\underset{\sim}{x})$ to a fixed Cartesian coordinate system. With these conventions, the interior of the body can be identified with an open set Ω of $\mathbb{R}^N$ ($N = 2$ or 3).

The body is subjected to body forces of intensity $\underset{\sim}{f}$ per unit volume in the reference configuration and to surface tractions $\underset{\sim}{g}$, measured per unit area in the reference configuration, prescribed on a portion Γ_2 of the boundary Γ of Ω. Both $\underset{\sim}{f}$ and $\underset{\sim}{g}$ might depend on the displacement field $\underset{\sim}{u}$. This displacement field takes on prescribed values $\underset{\sim}{u}_0$ on a portion Γ_1 of Γ and we have :

$$\Gamma = \overline{\Gamma}_1 \cup \overline{\Gamma}_2 \ , \ \Gamma_1 \cap \Gamma_2 = \phi .$$

Writing the laws of force and moment balance in Lagrangian coordinates, and given a constitutive law of hyperelastic type on Ω, we can characterize formally the equilibrium positions of the considered body as the solutions of the following system :

Laws of force and moment balance :

(5.1) $$-\underset{\sim}{\mathrm{DIV}}_R \ \underset{\sim}{T}_R(\underset{\sim}{x}) = \underset{\sim}{f} \ \underline{\text{in}} \ \Omega, \ \underset{\sim}{T}_R \cdot \underset{\sim}{\nu} = \underset{\sim}{g} \text{ on } \Gamma_2 \ ;$$

Hyperelastic constitutive law :

(5.2) $$\underset{\sim}{T}_R = \frac{\partial \mathcal{W}}{\partial \underset{\sim}{\nabla u}}[\underset{\sim}{x}, \underset{\sim}{I}+\underset{\sim}{\nabla u}, \mathrm{adj}(\underset{\sim}{I}+\underset{\sim}{\nabla u}), \det(\underset{\sim}{I}+\underset{\sim}{\nabla u})] \ ;$$

Orientation preservation :

(5.3) $$\det(\underset{\sim}{I}+\underset{\sim}{\nabla u}) > 0 \ \underline{\text{a.e. in}} \ \Omega \ ;$$

Boundary condition on $\underset{\sim}{u}$:

(5.4) $$\underset{\sim}{u} = \underset{\sim}{u}_0 \ \underline{\text{on}} \ \Gamma_1 .$$

Above, $\underset{\sim}{T}_R$ denotes the first Piola-Kirchhoff stress-tensor which characterizes the contact forces in the present configuration which are applied through an elementary surface, defined in the reference configuration, by one half on the body on the remainder ([15]). Moreover, $\underset{\sim}{\nu}$ represents the unit external normal vector to the body before deformation and $\mathcal{W}$ represents the elastic stored-energy density, measured per unit volume of the reference configuration. Typically, this stored-energy function can be decomposed into the sum:

(5.5) $$\mathcal{W}(\underset{\sim}{x}, \underset{\sim}{F}, \underset{\sim}{G}, \delta) = \mathcal{G}_1(\underset{\sim}{x}, \underset{\sim}{F}) + \mathcal{G}_2(\underset{\sim}{x}, \underset{\sim}{F}, \underset{\sim}{G}, \delta)$$

where $\mathcal{G}_1$ is a regular convex real function defined on $\Omega \times \mathbb{R}^{N\times N}$ and where $\mathcal{G}_2$, from $\Omega \times \mathbb{R}^{N\times N} \times \mathbb{R}^{N\times N} \times \mathbb{R}_+^{\star}$ into $\overline{\mathbb{R}}$, can be singular for zero values of the determinant δ. For example, for OGDEN's type materials, we may have [16] :

(5.6) $$\mathcal{W}(\underset{\sim}{x}, \underset{\sim}{F}, \underset{\sim}{G}, \delta) = C_1 |\underset{\sim}{F}|^2 + C_2 |\underset{\sim}{G}|^2 + C_3 \delta^2 - C_4 \ln(\delta) .$$

For a constitutive law given by (5.2) and (5.5), after elimination of $\underset{\sim}{T}_R$ between the law of force balance (5.1) and the constitutive law (5.2), we may formulate the equilibrium system (5.1)-(5.4) as the following variational system.

<u>Find</u> $\underset{\sim}{u} \in \{U^{1,s} + \underset{\sim}{u}_0\}$, $\underset{\sim}{A} \in (L^{s^\star}(\Omega))^{N\times N}$, <u>such that</u> :

(5.7) $$\delta = \det(\underset{\sim}{I}+\underset{\sim}{\nabla u}) > 0, \ \underline{\text{a.e. in}} \ \Omega \ ;$$

(5.8) $$\int_\Omega \{\frac{\partial \mathcal{G}_1}{\partial \underset{\sim}{F}}(\underset{\sim}{x}, \underset{\sim}{I}+\underset{\sim}{\nabla u}) \cdot \underset{\sim}{\nabla v} + \underset{\sim}{A} \cdot \underset{\sim}{\nabla v}\} d\underset{\sim}{x} = \int_\Omega \underset{\sim}{f} \cdot \underset{\sim}{v} \, d\underset{\sim}{x} + \int_{\Gamma_2} \underset{\sim}{g} \cdot \underset{\sim}{v} \, d \ , \quad \forall \underset{\sim}{v} \in U^{1,s} \ ;$$

$$(5.9)\quad \int_\Omega \underset{\sim}{A}\cdot\underset{\sim}{E}\,d\underset{\sim}{x} = \int_\Omega \frac{\partial \mathcal{G}_2}{\partial \underset{\sim}{G}}\left(\underset{\sim}{x},\underset{\sim}{I}+\underset{\sim}{\nabla}\underset{\sim}{u},\operatorname{adj}(\underset{\sim}{I}+\underset{\sim}{\nabla}\underset{\sim}{u}),\delta\right)\cdot\frac{\partial \operatorname{adj}(\underset{\sim}{I}+\underset{\sim}{\nabla}\underset{\sim}{u})}{\partial \underset{\sim}{\nabla}\underset{\sim}{u}}\cdot\underset{\sim}{E}\,d\underset{\sim}{x}\,\cdot$$
$$+\int_\Omega \frac{\partial \mathcal{G}_2}{\partial \underset{\sim}{F}}\left(\underset{\sim}{x},\underset{\sim}{I}+\underset{\sim}{\nabla}\underset{\sim}{u},\operatorname{adj}(\underset{\sim}{I}+\underset{\sim}{\nabla}\underset{\sim}{u}),\delta\right)\cdot\underset{\sim}{E}\,d\underset{\sim}{x}\,+$$
$$+\int_\Omega \frac{\partial \mathcal{G}_2}{\partial \delta}\left(\underset{\sim}{x},\underset{\sim}{I}+\underset{\sim}{\nabla}\underset{\sim}{u},\operatorname{adj}(\underset{\sim}{I}+\underset{\sim}{\nabla}\underset{\sim}{u}),\delta\right)\cdot\frac{\partial \det(\underset{\sim}{I}+\underset{\sim}{\nabla}\underset{\sim}{u})}{\partial \underset{\sim}{\nabla}\underset{\sim}{u}}\cdot\underset{\sim}{E}\,d\underset{\sim}{x},\quad \forall \underset{\sim}{E}\in\left(L^s(\Omega)\right)^{N\times N}.$$

The notation $\operatorname{adj}(\underset{\sim}{F})$ represents the adjugate of matrix F i.e. the transpose of the cofactor matrix. The usual notations are used for the Sobolev spaces $L^s(\Omega)$, $L^{s^\star}(\Omega)$, $(ss^\star = s + s^\star)$ and $W^{1,s}(\Omega)$. We have also

$$(5.10)\qquad U^{1,s} = \{\underset{\sim}{v}\in\underset{\sim}{W}^{1,s}(\Omega),\ \underset{\sim}{v}=0 \text{ on } \Gamma_1\}$$

The exponent s is related to the energy function $\mathcal{W}$ by continuity and coercivity requirements on $\mathcal{W}$.

The variational formulation (5.7)-(5.9) is unusual in finite elasticity, where $\underset{\sim}{A}$ is often eliminated between (5.7) and (5.9). Nevertheless, this decomposed formulation (5.7)-(5.9) is very interesting because it is close from the variational formulation of equilibrium problems in *incompressible hyperelasticity*. Under that form, the augmented lagrangian decomposition described in Sec. 4, already introduced in [12] for incompressible hyperelasticity, can be easily generalized to compressible hyperelasticity.

5.3. <u>Augmented lagrangian decomposition of (5.7)-(5.9)</u>. The extra variable needed for the decomposition of the equilibrium equations appears quite naturally to be the deformation gradient matrix :

$$(5.11)\qquad \underset{\sim}{F} = \underset{\sim}{I} + \underset{\sim}{\nabla}\underset{\sim}{u},\quad (\underset{\sim}{I} = \text{identity matrix on } \mathbb{R}^{N\times N}).$$

The augmented lagrangian $\mathcal{L}$ associated to our problem is then (see (4.8)) :

$$(5.12)\quad \begin{cases} \mathcal{L}(\underset{\sim}{v},\underset{\sim}{H},\underset{\sim}{\mu}) = \displaystyle\int_\Omega \mathcal{G}_2(\underset{\sim}{x},\underset{\sim}{H},\operatorname{adj}\underset{\sim}{H},\det\underset{\sim}{H})\,d\underset{\sim}{x} + \int_\Omega \mathcal{G}_1(\underset{\sim}{x},\underset{\sim}{I}+\underset{\sim}{\nabla}\underset{\sim}{v})\,d\underset{\sim}{x} \\ \qquad + \displaystyle\int_\Omega \frac{R}{2}|\underset{\sim}{I}+\underset{\sim}{\nabla}\underset{\sim}{v}-\underset{\sim}{H}|^2 d\underset{\sim}{x} - \int_\Omega \underset{\sim}{\mu}\cdot(\underset{\sim}{I}+\underset{\sim}{\nabla}\underset{\sim}{v}-\underset{\sim}{H})\,d\underset{\sim}{x}, \end{cases}$$

R being an arbitrary strictly positive constant. Now, as in Sec. 4, the variational system (5.7)-(5.9) of equilibrium equations can be written equivalently as the Lagrangien system below :

<u>Find</u> $\{\underset{\sim}{u},\underset{\sim}{F},\underset{\sim}{\lambda}\}\in(U^{1,s}+\underset{\sim}{u}_0)\times Y\times\left(L^{s^\star}(\Omega)\right)^{N\times N}$ <u>such that</u>

$$(5.13)\qquad \frac{\partial\mathcal{L}}{\partial\underset{\sim}{u}}(\underset{\sim}{u},\underset{\sim}{F},\underset{\sim}{\lambda})\cdot\underset{\sim}{v} = \int_\Omega \underset{\sim}{f}\cdot\underset{\sim}{v}\,d\underset{\sim}{x} + \int_{\Gamma_2}\underset{\sim}{g}\cdot\underset{\sim}{v}\,d\ ,\quad \forall\underset{\sim}{v}\in U^{1,s}\ ;$$

$$(5.14)\qquad \frac{\partial\mathcal{L}}{\partial\underset{\sim}{F}}(\underset{\sim}{u},\underset{\sim}{F},\underset{\sim}{\lambda})\cdot\underset{\sim}{H} = 0,\quad \forall\underset{\sim}{H}\in\left(L^s(\Omega)\right)^{N\times N}\ ;$$

$$(5.15)\qquad \frac{\partial\mathcal{L}}{\partial\underset{\sim}{\lambda}}(\underset{\sim}{u},\underset{\sim}{F},\underset{\sim}{\lambda})\cdot\underset{\sim}{\mu} = 0,\quad \forall\underset{\sim}{\mu}\in\left(L^{s^\star}(\Omega)\right)^{N\times N}.$$

Above Y denotes the set of elements of $\left(L^s(\Omega)\right)^{N\times N}$ with a strictly positive determinant almost everywhere in Ω.

5.4. Solution algorithm for the lagrangian system (5.13)-(5.15). Applying
the Uzawa algorithm (4.12)-(4.16) to the solution of the augmented lagrangian formulation (5.13)-(5.15) of the equilibrium equations leads to the
following sequence of operations :

(5.16) $\{\lambda^0, \underline{\mu}^{-1}\} \in \left(L^{s^\star}(\Omega)\right)^{N \times N} \times \{U^{1,s} + \underline{\mu}_0\}$ *given* ;

then, for $n \geqslant 0$, $\underline{u}^{n-1}$ *and* λ^n *being known, we compute* $\{\underline{u}^n, \underline{F}^n\}$ *by block rela-*
xation, i.e. by

(5.17) *setting* $\underline{u}_0^n = \underline{u}^{n-1}$,

and by computing sequentially $\underline{u}_k^n$ *and* $\underline{F}_k^n$ *by solving :*

(5.18) $\frac{\partial \mathcal{L}}{\partial \underline{F}}(\underline{u}_{k-1}^n, \underline{F}_k^n, \lambda^n) \cdot \underline{H} = 0, \quad \forall \underline{H} \in \left(L^s(\Omega)\right)^{N \times N}$,

(5.19) $\int_\Omega \{\frac{\partial \mathcal{G}_1}{\partial \nabla \underline{u}}(\underline{x}, \underline{I} + \nabla \underline{u}_k^n) + R(\underline{I} + \nabla \underline{u}_k^n - \underline{F}_k^n) - \lambda^n\} \cdot \nabla \underline{v} \, d\underline{x} = \int_\Omega \underline{f} \cdot \underline{v} \, d\underline{x} + \int_{\Gamma_2} \underline{g} \cdot \underline{v} \, ds$,
   for any $\underline{v}$ *in* $U^{1,s}$;

once $\{\underline{u}^n, \underline{F}^n\}$ *known,* λ^n *is updated by*

(5.20) $\lambda^{n+1} = \lambda^n - \mathcal{R}_z [R(\underline{I} + \nabla \underline{u}^n - \underline{F}^n)]$.

In the numerical applications, since we are always working with finite di-
mensional approximations of $\left(L^s(\Omega)\right)^{N \times N}$, the Riesz mapping $\mathcal{R}_z$ is replaced by
the identity mapping. The updating of λ^n by (5.20) is then completely
straight-forward and the whole algorithm above reduces the solution of the
equilibrium equations in compressible hyperelasticity to a sequence of quadratic problems (5.19) in displacements and of local problems (5.18) in
deformation gradients.

Since $\mathcal{G}_1$ only corresponds to the convex part of the elastic energy
function $\mathcal{W}$, the displacement problem (5.19) formally corresponds to an
unconstrained uniformly convex minimization problem, set on the linear
space $U^{1,s}$. Among all the solution methods existing for such problems, we
have picked a conjugate gradient method with preconditioning of incomplete
Choleski type (I C C G algorithm [17]). The preconditioning matrix is taken
very sparse, symmetric positive definite, and invariant during the iteration process. Its inversion will therefore be very cheap even in the
case of large three-dimensional finite element approximations of $U^{1,s}$. Due
to the convexity of (5.19), and since the solution $\underline{u}_{k-1}^n$ at previous step
usually is a good approximation of its final solution $\underline{u}_k^n$, the conjugate
gradient algorithm will converge quickly. Nevertheless, special attention
must be paid to the choice of the preconditioning matrix in order to avoid
unnecessary oscillations in the iterative process (5.16)-(5.20).

5.5. Three-dimensional study of the local problem (5.18) in deformation gradient. If $\left(L^s(\Omega)\right)^{N \times N}$ is approximated by a space of piecewise constant functions, problem (5.18), which does not involve any spatial derivative of $\underline{F}$, reduces to a sequence of independent local problems. For R sufficiently large, one can prove in addition that each local problem is equivalent to ([18]) :

(5.21) $\underset{\underline{F} \in Y_{loc}}{\text{Min}} \quad J(\underline{F}) = \mathcal{G}_2(\underline{x},\underline{F},\text{adj}\,\underline{F},\det\,\underline{F}) + \dfrac{R}{2}\left|\underline{F}-\underline{I}-\underline{\nabla}\underline{u}_{k-1}^n\right|^2 + \underline{\lambda}^n \cdot \underline{F}.$

with

(5.22) $Y_{loc} = \left\{\underline{F} \in \mathbb{R}^{N \times N}, \ \det\,\underline{F} > 0\right\}.$

If, as it is generally the case, $\mathcal{G}_2$ takes infinite values for negative determinant values of $\det\,\underline{F}$, Y_{loc} can be as well replaced by $\mathbb{R}^{N \times N}$. Due to its small dimension, this problem (5.21) could be solved by classical minimization techniques of multidimensional algebraic functions. But in fact, due to its local structure, which involves $\underline{F}$, its adjugate and its determinant, (5.21) can be reduced to a one-dimensional minimization problem for $N = 2$ or to a sequence of one-dimensional convex minimization problems for $N = 3$. From now on, we restrict ourselves to the most difficult case $N = 3$ (three dimensional structures).

The *decomposition* of the local minimization problem (5.21) for $N = 3$ is again based on *augmented lagrangian techniques*. For that purpose, we introduce three new variables $\underline{f} \in R^{36}$, $\underline{g} \in \mathbb{R}^9$ and $\underline{c} \in \mathbb{R}^9$ which permit a simple expression of $\text{adj}(\underline{F})$ and $\det(\underline{F})$ as functions of $\underline{F}$ and which are defined by :

$\sqrt{2}\,f_1 = F_5 + F_9$	$\sqrt{2}\,f_{13} = (F_8 + F_3)$	$\sqrt{2}\,f_{25} = (F_2 + F_6)$
$\sqrt{2}\,f_2 = F_5 - F_9$	$\sqrt{2}\,f_{14} = (F_8 - F_3)$	$\sqrt{2}\,f_{26} = (F_2 - F_6)$
$\sqrt{2}\,f_3 = F_6 + F_8$	$\sqrt{2}\,f_{15} = (F_9 + F_2)$	$\sqrt{2}\,f_{27} = (F_3 + F_5)$
$\sqrt{2}\,f_4 = F_6 - F_8$	$\sqrt{2}\,f_{16} = (F_9 - F_2)$	$\sqrt{2}\,f_{28} = (F_3 - F_5)$
$\sqrt{2}\,f_5 = F_6 + F_7$	$\sqrt{2}\,f_{17} = (F_9 + F_1)$	$\sqrt{2}\,f_{29} = (F_3 + F_4)$
$\sqrt{2}\,f_6 = F_6 - F_7$	$\sqrt{2}\,f_{18} = (F_9 - F_1)$	$\sqrt{2}\,f_{30} = (F_3 - F_4)$
$\sqrt{2}\,f_7 = F_4 + F_9$	$\sqrt{2}\,f_{19} = (F_7 + F_3)$	$\sqrt{2}\,f_{31} = (F_1 + F_6)$
$\sqrt{2}\,f_8 = F_4 - F_9$	$\sqrt{2}\,f_{20} = (F_7 - F_3)$	$\sqrt{2}\,f_{32} = (F_1 - F_6)$
$\sqrt{2}\,f_9 = F_4 + F_8$	$\sqrt{2}\,f_{21} = (F_7 + F_2)$	$\sqrt{2}\,f_{33} = (F_1 + F_5)$
$\sqrt{2}\,f_{10} = F_4 - F_8$	$\sqrt{2}\,f_{22} = (F_7 - F_2)$	$\sqrt{2}\,f_{34} = (F_1 - F_5)$
$\sqrt{2}\,f_{11} = F_5 + F_7$	$\sqrt{2}\,f_{23} = (F_8 + F_1)$	$\sqrt{2}\,f_{35} = (F_2 + F_4)$
$\sqrt{2}\,f_{12} = F_5 - F_7$	$\sqrt{2}\,f_{24} = (F_8 - F_1)$	$\sqrt{2}\,f_{36} = (F_2 - F_4)$

Table 5.1

(5.23) $$\underset{\sim}{f} = \underset{\sim}{T}\,\underset{\sim}{F} \quad (\underset{\sim}{T} \text{ given by Table 5.1}) ;$$

(5.24) $$g_j = \frac{1}{2}\varepsilon_i f^2_{4j-4+i}, \quad \forall j = 1,9, \; i = 1,4, \; \varepsilon_1 = -\varepsilon_2 = -\varepsilon_3 = \varepsilon_4 = 1 ;$$

(5.25) $$\underset{\sim}{G} = \underset{\sim}{g}.$$

With these new variables, it is easy to calculate

$$\text{adj}\,\underset{\sim}{F} = \underset{\sim}{G}^T, \quad \det \underset{\sim}{F} = F\cdot\underset{\sim}{G}/3.$$

Let us now introduce the following local augmented lagrangian :

(5.26) $$\begin{aligned}\mathcal{L}_\ell(\{\underset{\sim}{F},\underset{\sim}{G}\},\{\underset{\sim}{f},\underset{\sim}{g}\},\{\underset{\sim}{z},\underset{\sim}{t}\}) &= \mathcal{G}_2(\underset{\sim}{x},\underset{\sim}{F},\underset{\sim}{G},\underset{\sim}{F}\cdot\underset{\sim}{G}/3) + \frac{R}{2}|\underset{\sim}{F}-\underset{\sim}{I}-\nabla \underset{\sim}{u}^n_{k-1}|^2 \\ &+ \underset{\sim}{\lambda}^n\cdot\underset{\sim}{F} + \frac{r}{2}\{|\underset{\sim}{f}-\underset{\sim}{T}\underset{\sim}{F}|^2 + |\underset{\sim}{g}-\underset{\sim}{G}|^2\} - \underset{\sim}{z}\cdot(\underset{\sim}{f}-\underset{\sim}{T}\underset{\sim}{F}) - \underset{\sim}{t}\cdot(\underset{\sim}{g}-\underset{\sim}{G}),\end{aligned}$$

r being an arbitrary strictly positive constant. The local minimization problem (5.21) can then be decomposed into the equivalent system :

(5.27) $$\begin{cases} \underline{\text{Find}}\ \{\{\underset{\sim}{F},\underset{\sim}{G}\},\{\underset{\sim}{f},\underset{\sim}{g}\},\{\underset{\sim}{z},\underset{\sim}{t}\}\} \in X_\ell, Y_\ell, Z_\ell\ \underline{\text{such that}} \\ \{\underset{\sim}{F},\underset{\sim}{G}\}\ \underline{\text{minimizes}}\ \mathcal{L}_\ell(\{\cdot,\cdot\},\{\underset{\sim}{f},\underset{\sim}{g}\},\{\underset{\sim}{z},\underset{\sim}{t}\})\ \underline{\text{over}}\ X_\ell, \\ \{\underset{\sim}{f},\underset{\sim}{g}\}\ \underline{\text{minimizes}}\ \mathcal{L}_\ell(\{\underset{\sim}{F},\underset{\sim}{G}\},\{\cdot,\cdot\},\{\underset{\sim}{z},\underset{\sim}{t}\})\ \underline{\text{over}}\ Y_\ell, \\ \{z,t\}\ \underline{\text{maximizes}}\ \mathcal{L}_\ell(\{F,G\},\{f,g\},\{\cdot,\cdot\})\ \underline{\text{over}}\ Z_\ell. \end{cases}$$

the sets X_ℓ, Y_ℓ, Z_ℓ being defined by :

$$X_\ell = \mathbb{R}^9 \times \mathbb{R}^9, \quad Z_\ell = \mathbb{R}^{36} \times \mathbb{R}^9,$$

$$Y_\ell = \left\{\{\underset{\sim}{f},\underset{\sim}{g}\} \in Z_\ell,\ g_j = \frac{1}{2}\varepsilon_i f_{4j+i-4}\right\}.$$

The solution of the local minimization problem (5.21) finally reduces to the iterative solution of its augmented lagrangian formulation (5.27) by the Uzawa and block relaxation algorithm (4.12)-(4.16). The variables $u,\underset{\sim}{F},\underset{\sim}{\lambda}$ are now respectively $\{\underset{\sim}{F},\underset{\sim}{G}\},\{\underset{\sim}{f},\underset{\sim}{g}\}$ and $\{\underset{\sim}{z},\underset{\sim}{t}\}$. In this algorithm, two elementary subproblems appear :

(5.28) $$\underset{\{\underset{\sim}{F},\underset{\sim}{G}\}\in X_\ell}{\text{Min}}\ \mathcal{L}_\ell(\underset{\sim}{F},\underset{\sim}{G},\underset{\sim}{f}^m_{\ell-1},\underset{\sim}{g}^m_{\ell-1},\underset{\sim}{z}^m,\underset{\sim}{t}^m),$$

(5.29) $$\underset{\{\underset{\sim}{f},\underset{\sim}{g}\}\in Y_\ell}{\text{Min}}\ \mathcal{L}_\ell(\underset{\sim}{F}^m_\ell,\underset{\sim}{G}^m_\ell,\underset{\sim}{f},\underset{\sim}{g},\underset{\sim}{z}^m,\underset{\sim}{t}^m).$$

Problem (5.29) is similar to (5.28) but simpler. That's why we only detail below the solution of (5.28), referring to [18] for the solution of (5.29).

5.6. Solution of (5.28). In this paragraph, we will suppose that the stored energy function is of OGDEN's type and is given by ([16]) :

$$\mathcal{W}(\underset{\sim}{x},\underset{\sim}{F},\underset{\sim}{G},\delta) = C_1|\underset{\sim}{F}|^2 + C_2|\underset{\sim}{G}|^2 + C_3\delta^2 - C_4\ \ell n\,\delta .$$

For this stored energy function, the solution of (5.28) is achieved by making the following change of variables

(5.30) $$\underset{\sim}{U} = \underset{\sim}{F} + \beta\underset{\sim}{G}, \quad \underset{\sim}{V} = \underset{\sim}{F} - \beta\underset{\sim}{G},$$

with

(5.31) $$\alpha = R + 4r, \quad \beta = \left((2C_2+r)/(R+4r)\right)^{\frac{1}{2}}.$$

Problem (5.28) is then transformed into :

(5.32) $$\underset{\{\underset{\sim}{U},\underset{\sim}{V}\}\in\mathbb{R}^{18}}{\text{Min}} \; j(\underset{\sim}{U},\underset{\sim}{V}) = \frac{\alpha}{4}(|\underset{\sim}{U}-\underset{\sim}{A}|^2 + |\underset{\sim}{V}-\underset{\sim}{B}|^2) + C_3 q^2 - C_4 \; \ell n(q),$$

with

(5.33) $$q = \underset{\sim}{F}\cdot\underset{\sim}{G}/3 = (|\underset{\sim}{U}|^2 - |\underset{\sim}{V}|^2)/12\beta,$$

(5.34) $$\underset{\sim}{A} = \frac{1}{2}\{R(\underset{\sim}{\nabla u}+\underset{\sim}{I}) - \underset{\sim}{\lambda} + \underset{\sim}{T}^t(r\; \underset{\sim}{f}^m_{\ell-1} - \underset{\sim}{z}^m) + (r\; \underset{\sim}{g}^m_{\ell-1} - \underset{\sim}{t}^m)/\beta\},$$

(5.35) $$\underset{\sim}{B} = \frac{1}{2}\{R(\underset{\sim}{\nabla u}+\underset{\sim}{I}) - \underset{\sim}{\lambda} + \underset{\sim}{T}^T(r\; \underset{\sim}{f}^m_{\ell-1} - \underset{\sim}{z}^m) - (r\; \underset{\sim}{g}^m_{\ell-1} - \underset{\sim}{t}^m)/\beta\}.$$

The solution of (5.32) is very easy to compute and is given by

(5.36) $$\underset{\sim}{U} = \underset{\sim}{A}/\left(\frac{\alpha}{2} + p/6\beta\right), \quad \underset{\sim}{V} = \underset{\sim}{B}/\left(\frac{\alpha}{2} - p/6\beta\right)$$

where p is the only root on the interval $(-3\alpha\beta, 3\alpha\beta)$ of the equation

(5.37) $$|\underset{\sim}{A}|^2/\left(\frac{\alpha}{2} + p/6\beta\right)^2 - |\underset{\sim}{B}|^2/\left(\frac{\alpha}{2} - p/6\beta\right)^2 - \frac{3\beta}{C_3}[p + \sqrt{p^2+8C_4C_3}] = 0.$$

The numerical solution of (5.28) is thus simply obtained by

(i) solving (5.37) by a Newton's method,

(ii) computing $\{\underset{\sim}{U},\underset{\sim}{V}\}$ by (5.36),

(iii) computing $\{\underset{\sim}{F},\underset{\sim}{G}\}$ by inverting (5.30).

5.7. Axisymmetric numerical experiments. The formulation of the equilibrium equations and the treatment of the local problem in deformation gradient are given in [12] for the case of axisymmetric loadings of axisymmetric hyperelastic incompressible bodies. For the numerical solution of such problems, we choose here finite element approximations of the displacement space $U^{1,s}$ and of the deformation gradient space $\left(L^s(\Omega)\right)^5$, based on the 4 nodes asymmetric finite element developped by RUAS [19]. The element's geometry is a triangle, the degrees of freedom in displacement are their values at each vertex and at the midpoint of one side, the degrees of freedom in deformation gradient are their values at the center.

Assembling these elements three by three, we obtain seven nodes symmetric finite superelements. Approximate displacements are taken continuous at element interfaces, approximate deformation gradients are not (they are piecewise constants).

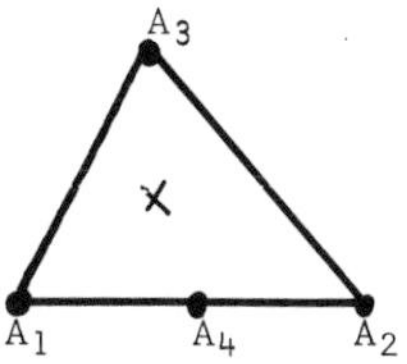

the asymmetric finite element

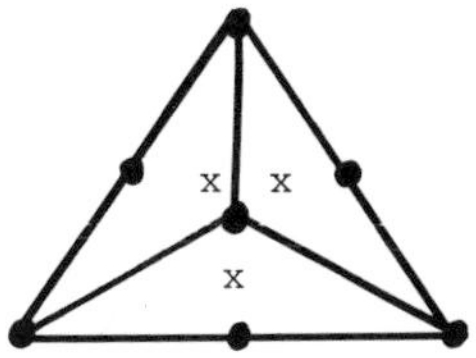

symmetric assembly of 3 elements

* degrees of freedom in deformation gradients,
• degrees of freedom in displacements.

Fig. 5.1

The numerical problem to be solved corresponds then to the same a-grangien system (5.13)-(5.15), but set on the above finite element approximations of $U^{1,s}$ and $(L^s(\Omega))^{N\times N}$. Solution techniques remain unchanged compared to those described in the continuous case. The numerical example presented in this paragraph concerns the axial compression of an axisymmetric incompressible hyperelastic shaft whose shape is indicated below. For symmetry reasons, we restrict our domain Ω to the upper meridian section of this shaft. The mesh before and after compression is represented on the figure below. Observe the surface discontinuity on the shaft after deformation. Such a singularity is in complete agreement with the experimental studies and would be very difficult to obtain by usual numerical techniques.

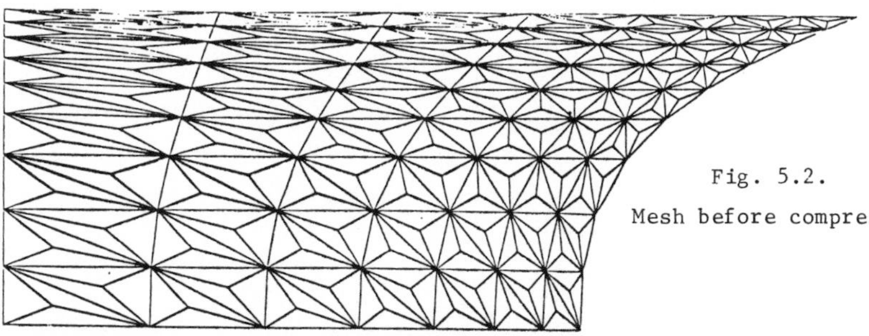

Fig. 5.2.
Mesh before compression

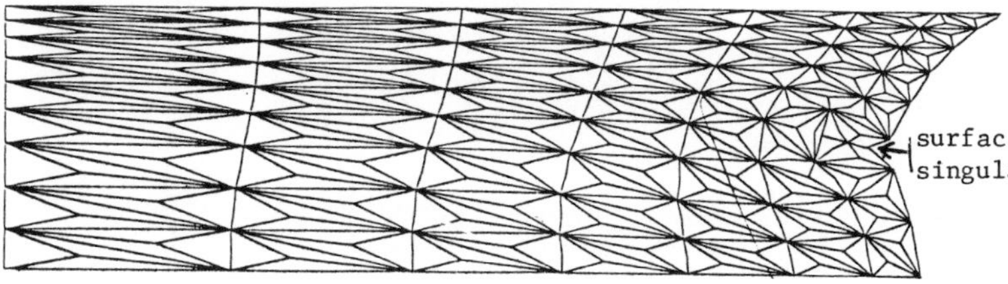

Fig. 5.3. 30% compression

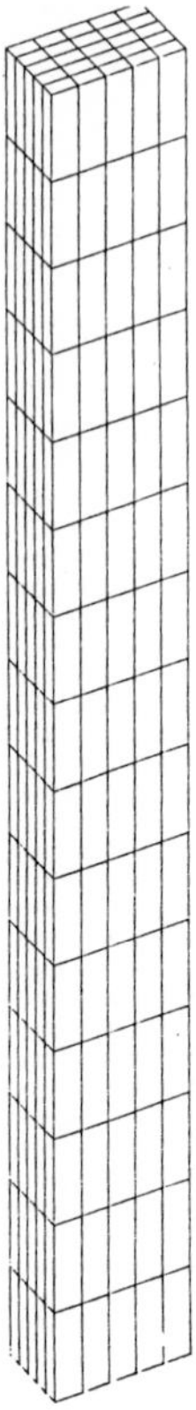

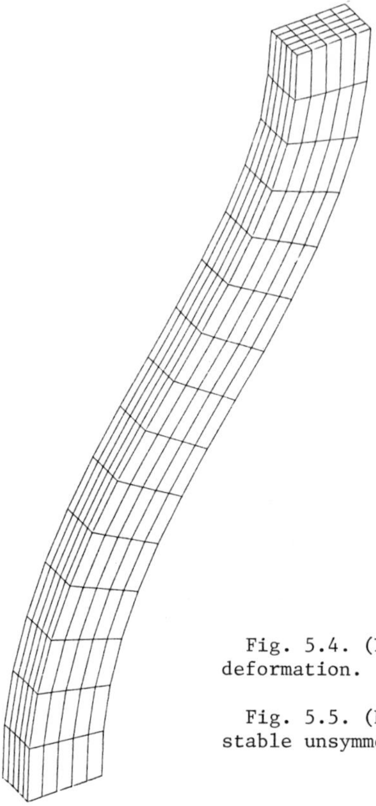

Fig. 5.4. (Left) mesh before deformation.

Fig. 5.5. (Right) 10% compression stable unsymmetric solution.

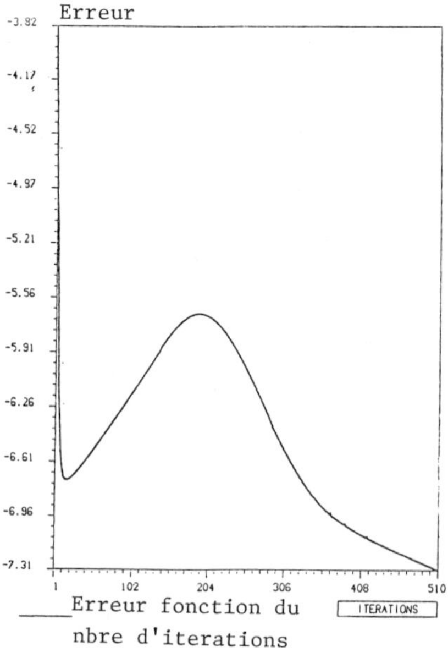

Fig. 5.6. Convergence rate as a function of the iteration number.

5.8. Three dimensional numerical experiments. The numerical example presented here illustrates the capability of the above numerical method for computing *stable postbuckling equilibrium positions* of hyperelastic bodies, even in three-dimensional configurations. The considered body is a 2×2×20 compressible elastic beam shortened to 90% of its initial length and subjected to a very small surface pressure on one of its faces. The stored energy function of the beam is supposed to be given by the function of (5.6).

The displacement space is approximated by standard isoparametric 8 nodes hexahedral elements (Q^1 cubes), the approximate deformation gradients being constant on each element. Two solutions are then obtained by using the augmented lagrangian techniques of this paragraph :

(i) an unstable symmetric solution with almost no horizontal displacements ;

(ii) a stable unsymmetric solution with large horizontal displacements.

The aspect of the beam before and after deformation is indicated on Fig. 5.4 and 5.5. For symmetry reasons, we only consider the upper part of the beam.

It is particularly interesting here to monitor the convergence of the Uzawa algorithm. Measuring the convergence rate by $\|\underset{\sim}{I}+\nabla\underset{\sim}{u}^n-\underset{\sim}{F}^n\|_{0,2}$ at each iteration, we observe that this convergence indicator first decreases while the computed solution $\underset{\sim}{u}^n$ goes from zero to the unstable symmetric solution, then increases as $\underset{\sim}{u}^n$ automatically leaves the neighborhood of this symmetric solution, and finally decreases towards zero as $\underset{\sim}{u}^n$ approaches the final stable buckled solution (see Fig. 5.6). This whole iterative process *goes on completely automatically* without any operator's action or incremental loading technique, *by the simple execution of algorithm* (5.16)-(5.20) with $\underset{\sim}{u}^0 = 0$.

REFERENCES

[1] GLOWINSKI, R.G.- Numerical Methods for Nonlinear Variational Problems. Springer Verlag, New York, 1983.

[2] ADAMS, R.A.- Sobolev Spaces. Academic Press, New York, 1975.

[3] POLAK, E.- Computational Methods in Optimization. Academic Press, New York, 1971.

[4] POWELL, M.J.D.- Restart procedure for the conjugate gradient method. Math. Programming, 12, (1977), pp. 148-162.

[5] GLOWINSKI, R. ; KELLER, H.B. ; REINHART, L.- Continuation - Conjugate Gradient methods for the least squares solution of nonlinear boundary value problems (to appear).

[6] REINHART, L.- On the numerical analysis of the Von Karman equations : Mixed Finite Element Approximation and Continuation Techniques. Numerische Math., 39, (1982), pp. 371-404.

[7] BRISTEAU, M.O.; GLOWINSKI, R. ; PERIAUX, J. ; PERRIER, P. ; PIRONNEAU,O. POIRIER, G.- Application of optimal control and finite element methods to the calculation of transonic flows and incompressible viscous flows, in Numerical Methods in APplied Fluid Dynamics, B. Hunt ed., Academic Press, London, 1980, pp. 203-312.

[8] TAYLOR, C. ; HOOD, P.- A numerical solution of the Navier-Stokes equations using the finite element technique. Computers and Fluids, 1, (1973), pp. 73-100.

[9] GLOWINSKI, R.G. ; PERIAUX, J. ; DINH, Q.V.- Domain Decomposition methods for nonlinear problems in Fluid Dynamics. Comp. Meth. in Appl. Mech. Eng. (to appear).

[10] GLOWINSKI, R.G. ; MAROCCO, A.- Sur l'approximation par éléments finis d'ordre un et la résolution par pénalisation - dualité d'une classe de problèmes de Dirichlet nonlinéaires. RAIRO série rouge, R2 (1975) p. 41-76.

[11] GLOWINSKI, R.G. ; FORTIN, M. (eds). Méthodes de Lagrangien Augmenté. Application à la Résolution Numérique des Problèmes aux Limites, Dunod, Paris, 1982.

[12] GLOWINSKI, R.G. ; LE TALLEC; P.- Numerical solution of problems in incompressible finite elasticity by augmented lagrangian methods (I). Two-dimensional and axisymmetric problems. SIAM J. Appl. Math. 42, (1982), pp. 400-429.

[13] GLOWINSKI, R.G. ; LE TALLEC, P.- Numerical solution of problems incompressible finite elasticity by augmented lagrangian methods (II). Three-dimensional problems.(to appear in SIAM J. Appl. Math.).

[14] GLOWINSKI, R.G. ; LE TALLEC, P.- Elasticité nonlinéaire. Formulation mixte et méthode numérique associée. In Computing Methods in Applied Sciences and Engineering. GLOWINSKI-LIONS ed. North-Holland, Amsterdam, 1982.

[15] HUGHES, T.J.R. ; MARSDEN, J.- Mathematical Foundations of Elasticity, Prentice-Hall, 1983.

[16] CIARLET, P.G. ; GEYMONAT.- Sur les lois de comportement en élasticité nonlinéaire compressible. C.R.A.S. (1982).

[17] MEIJERINK, J.A. ; VAN DE VORST, H.A.- An iterative solution method for linear systems of which the coefficient matrix is a symmetric M matrix. Math. of Comp. 31 (1977).

[18] LE TALLEC, P. ; VIDRASCU, M.- Une méthode numérique pour les problèmes d'équilibre de corps hyperélastiques compressibles en grandes déformations. Rapport sectoriel MA 1 du Laboratoire Central des Ponts et Chaussées (1983).

[19] RUAS, V.- Méthodes d'éléments finis en élasticité incompressible nonlinéaire. Thèse d'Etat. Université Pierre et Marie Curie (1982).

Roland Glowinski
Université Pierre et Marie Curie
Place Jussieu 75230 Paris Cedex 5
FRANCE and INRIA

Patrick Le Tallec
Laboratoire Central des Ponts et Chaussees
Service de Mathematiques
58 boulevard LEFEBVRE 75015 Paris FRANCE

FUTURE DIRECTIONS IN LARGE SCALE SCIENTIFIC COMPUTING

James M. Hyman

I. INTRODUCTION.

American scientists are deeply concerned that our leadership role is slipping in large-scale scientific computer hardware and software development.[1-3] The recently publicized national efforts in Japan, West Germany, France (Telematique Program), and Great Britain (Esprit Program) raise doubts about our ability to maintain our technological edge in computer hardware and software. Losing this edge would have serious repercussions on our economic and national security. We have maintained our leadership over the past two decades largely because we have excelled in developing scientific computer codes.

These codes are an indispensable tool in many modern science and engineering projects. Traditionally, they have been designed by scientists or engineers in the field that the code is intended to model. By combining expert knowledge in a particular field with intuition and experimental clues and by incorporating a modest experience in numerical analysis and software design, these programmers have often devised ingenious methods and written codes that work well, with varying degrees of accuracy.

The capabilities of any scientific computer code limit our understanding of the problem being modeled and, hence, our ability to conceive new ideas and research directions. More accuracy is needed as the engineering designs become

ISBN 0-12-546080-5

more detailed and operate closer to the thresholds of failure. Future applications in reactor safety design, controlled fusion, and other highly technical fields will require details and accuracies beyond the capabilities of the current codes and in some cases beyond current scientific knowledge. These codes are large because the problem to be solved is complex, not necessarily because the codes are unstructured. We cannot wait until all the modeling aspects are well understood before beginning developing a new code; rather, we must begin now to shorten future code development time and ease the process of extending and modifying existing software.

We can expect that the continuing revolution in micro-electronics will decrease the cost of computing and increase the speed and memory available, but these computer hardware improvements will allow present codes to be pushed only a limited amount beyond their current capabilities. Additional capabilities must come from increased accuracy in the physics models and more efficient numerical approximations to the models. Significant improvements in these areas are difficult to gain; experience has shown that even when advances are made they are often not fully exploited in some of the large physics production codes because these codes are too inflexible.

Because the architecture of the new machines will certainly be different from that of today's supercomputers we will have to re-examine traditional methods and identify those methods that utilize the inherent powers of the new machines. In the future, we must reduce the risk of locking our codes into a particular machine, a difficult task because methods tailored specifically for a particular machine architecture will probably become more the norm than the exception. We can, by keeping machine-dependent codes in libraries with standard user interfaces, strive to keep the user's scientific applications codes portable.

Many production-code programmers are scientists who would rather be concentrating on a few problems of long-term importance and who prefer to let interested experts do the support work. Often, these scientists are bogged down with day-to-day problems and find it tiresome when forced repeatedly to be ingenious in overcoming inadequacies in the computer system or the support software. When they are responsible for the physics as well as the numerics and the computer system's idiosyncrasies, they become less effective physicists or engineers. These extraneous, but necessary, duties gobble up their scarcest resource—time.

My goal in this report is to improve the effectiveness of applications programmers who currently use the computer but are not necessarily experts in scientific computing. I discuss those critical points in computer-structured design techniques that can improve the reliability, efficiency, and accuracy of future production codes. Also, I map out the steps we can take now to reduce the time and effort needed to build improved production codes in the future.

Much has been written on structured design techniques, and it is a familiar subject to most experienced programmers. Rather than repeat detailed information here, I will address the crucial software issues and the steps we must take for production codes to exploit fully the power of future computers. Software flexibility, modularity, reliability, efficiency, and restricted data flow and documentation are fundamental to the design of a new physics production code. In many computer installations, routines in the mathematics support library meet these standards. However, many of these libraries lack unification in standard quality-assurance procedures, documentation, data structures, naming conventions, and coding styles. Also, the libraries almost totally lack high-level mathematics routines to support large-scale scientific computing.

If these high-level routines existed, they could perform many of the common procedures found in physics production codes, including grid generation, rezoning, numerical interpolation, differentiation, and integration; they could approximate differential boundary conditions and solve large, sparse nonlinear systems of equations. The mathematical theory and software design for many of these routines have not been developed. Development requires that applications programmers, theoretical numerical analysts, and the scientific computing research communities in the national laboratories, industry, and the universities all work together.

II. INITIAL PLANNING STAGE.

Designing and implementing a new physics production code are costly and time consuming. The life cycle of a production code consists of several phases. First, we must prepare an initial needs analysis plan, deciding in detail what will be done with the results, what form they should appear in, how soon they are needed, what accuracy is required, and how many similar problems are to be solved. Next, the program specifications must be prescribed and used to determine the software design of the code. This design will be based on the underlying structure of the problem being solved, the basic algorithms being used, and the major data representations and their interrelationships. The project then moves from the design phase into the construction phases of implementation and a continuing maintenance program. In these phases, just as a construction company must wait for the architect's detailed plans before erecting walls and windows, so must the programmers wait for the software architect to complete the system design before significant pieces of the system are implemented.

The design phases are concerned with reliability, user friendliness, efficiency, flexibility, and other global properties of the code. Because these are global properties, we should plan for them before construction is begun; otherwise, they generally are not attainable. Thus, the structured design of a major code is primarily an exercise in a priori organization. Good organization is the key to a good program: if the parts are properly organized, fit together in a logical structure, and can be constructed in a rational sequence of intermediate stages, the program will almost write itself. To implement a large production code, the programming team must know exactly what the code will do and should know why it is being done in a particular way.

A. Management

Bad management can cause even the best designed codes to fail. For example, successful production code projects need strong support, and it is often difficult to convince managers that software, not hardware, will pose the most difficult problems and will account for the most impressive advances. The software design phase is time consuming, expensive, and necessary, but in many organizations, these expenditures are a touchy subject. With this in mind, I have the following advice for the code team managers: it is a humbling experience to make a multimillion-dollar mistake, such as funding a production code that never successfully solves a problem. When developing a major production code, errors in the project's design phase extract high penalities. For this reason, early in the design phase, simulate what you plan to do, repeat as many times as necessary for you to be confident in the final result, and also have a contingency plan for failures. When starting the initial design, think what could happen should, for example, key people leave or the budget be spent thereby stopping the project when 80% completed, and plan the project so that you can live with the outcome. Then, you can move ahead with confidence, unhampered by a sense of impending doom. This planning, essential to success even if the project is not completed, can be done in the design phase if you are sensitive to potential spinoffs.

B. Spinoffs

In both privatly-funded and government-funded projects, one of the best potential spinoffs is to extend the capabilities of existing codes; we must consider existing programs when designing a new production code. These routines, whose correctness and usefulness have been confirmed by actual service, are precious and should play an important role in future developments. To reduce duplications of effort, we must strive for compatible, evolutionary, not revolutionary, changes. However, you must set high, but realistic, goals and not allow these existing codes to remain static in the comfortable stagnation zone (that no-man's land where you feel too much at home).

Many production codes are developed under government contracts. When this is the case, upper management and the funding agencies should encourage spinoffs and reward the transferring of new software and numerical methods to related areas. These new developments can upgrade the research capabilities of industry, university computer centers, and the government laboratories. Upper management and the funding agencies should reward those persons who generate highly visible spinoffs, the tangible results of the technology developed under their sponsorship.

For this approach to be successful, the new software should be compatible with existing techniques and simple enough so that potential users can observe tangibly better results in a trial run than those existing methods can produce. Software packages that behave like the existing ones, but work better, are the most readily accepted.

III. MODELS, METHODS, AND MACHINES.

The physics model is the cornerstone of the production code, but in many important areas, good physics models (for example, turbulence models) do not exist. When they do exist, often only crude numerical methods are available to solve the models (such as methods for large, nonlinear sparse systems). Sometimes, even when reliable physics models and accurate, efficient numerical methods exist, our current computers are too small or too slow to solve them accurately (as is true in three-dimensional fluid dynamics). For problems where new technology must be developed, we should develop a code using what is known and support research in areas that will improve the capabilities of future codes.

Another problem that must be overcome is the schism that often exists between the physicist/applications programmers and theoretical numerical analysts. One group thinks of themselves as pioneers in a new and exciting practical field, and the other as generalists abstracting the simplicity of effective numerical schemes. For example, to find an algorithm that works, applied physics groups have been known to run a computer code through hundreds of tests in an Edisonian approach, rather than trying to understand mathematically what is going on and building from there. Because of the expense and complexity of future problems, theoretical research must be used more than ever to guide us in finding more reliable algorithms. On the other hand, theoreticians must present their results in a form that the applications programmers can test in situations far more complex than solid theoretical analysis can justify.

Currently, many questions are not mathematically tractable, but the solution, nevertheless, is required. When intractability is the case, the code cannot be built on first principles. For the results of such codes to agree with the experimental results, heuristic knobs must be introduced to account for inaccuracies in the physics model or the numerics. Good applications programmers always avoid introducing a knob that affects well-understood physical models because, when using a design code, the modelers risk blurring the codes and experiments together, which (unless the code is free of knobs) can be a handicap to understanding the physics correctly. If we are to go beyond modeling, which interpolates between experimental results, to the extrapolation of such results and the simulation of new designs, then we must continue to develop and implement better models based on first principles and reduce the number of knobs.

For these new models to affect a design, they must be in the design code. It is unrealistic to expect the physics model, the numerical methods, or the computer architecture to be stable during the lifetime of the code. Any up-to-date large production program will eventually exist in a multitude of different versions. It is common sense to leave options open so we can anticipate and profit from advances whenever possible. Thus, flexibility must be a design criteron.

A. Flexibility

Typically, production codes are used for a long time without major changes in their structure. Major revisions are rarely encouraged and often not allowed because extensive revalidation will be needed before the results can be used for engineering purposes. In the design state, a code that is more flexible than necessary is far better than one that is inflexible and unable to handle future problems. Also, underestimating the size of the physics model necessary to correctly approximate a problem is more common than is overestimating it. If a code is inflexible, then implementing new models or methods or converting the code to a new machine is always risky and costly and will be put off as long as possible. If the code is flexible, then it will evolve, and we can often avoid these traumas completely—or at least experience them less often.

To predict when a model or method will be useful in more complicated situations, we must be able to test the idea easily. The best approach depends largely on the scale of the problem; the best approach for a simple test problem may be unacceptable for the production runs. Flexibility is the key to coping with this scale-up problem.

Much of the programming in the current generation of production codes was done by the code developers. We who worked on these codes remember it was often a lot of fun—but sometimes the magnificent dreams turned into nightmares: the communication links between different parts of the code grew out of control; capabilities that we expected to be easily achieved turned out to be extremely difficult or impossible; the program grew into a tangled skein of entwined coding; and the fact that "it works today" became its strongest point.

To prevent such mishaps, we need a means of concentrating on one part of the code at a time. Only through forethought and planning can we avoid spawning a monster code. One approach is to develop the code around a few simple structures and reduce the number of interconnections that the programmer must be aware of. We can accomplish this by dividing the problem into a set of easy-to-understand code modules.

B. Modularity

A module is a coherent package of codes that can be programmed, verified, and documented independently of the applications code. The interfaces between a module and the applications code should be simple and transparent so that changes can be made with minimal effect on the other parts of the system. The module concept allows us to divide a large problem into small, easily understood pieces that can be programmed separately and verified at each step of the development process. Later, these pieces will be assembled to yield a solution to the original problem.

The success of the modular approach depends on how well the modules are chosen. A common function that is not identified early enough may be duplicated in many different modules and obscure the module's purpose. On the other hand, modularity can be carried too far. Beyond a certain point, dividing the system into seldom-used pieces with only a few lines of code will introduce additional complexities, causing the overall design to be cumbersome. Care must be taken in the design stage to prevent either of these mishaps.

Applications programmers can understand what a module does and then use it without knowing what goes on inside. But before they do, they must have confidence in the reliability and efficiency of the numerical methods and the allowed data flow. Not only must the module produce correct results in a timely fashion, but it must do so without influencing the rest of the code. To insure some safety for the user, high standards for the modules must be a design criterion.

The quality assurance standards that insure the reliability, efficiency, restricted minimal data communication links, and documentation of a module of code are described in Appendix A. These standards are necessary because a program cannot be made fully reliable by testing. Reliability must be built into the program by applying a systematic approach based on a high degree of precision and structure. Using this approach programmers can write reliable programs and convince themselves and others that their programs are correct.

Also, we must make the modules friendly and easy to use so that applications programmers are tempted to use them rather than to write their own programs or use a less reliable or less efficient routine. Meeting these goals requires both good documentation and a good support staff for the programmer.

IV. THE LIBRARY CONCEPT.

In many computer installations, routines in the mathematical libraries meet the reliability, efficiency, data flow, and documentation standards advocated in Appendix A. These libraries often include routines from the International Mathematical and Statistical Library (IMSL); the Numerical Algorithms Group (NAG) at Oxford; the National Activity to Test Software (NATS) effort, including EISPACK and FUNPACK and their spinoffs, LINPACK and MINPACK; and the national laboratory SLATEC mathematics library.

Also, the mathematics software literature in journals, such as the ACM Transactions on Mathematical Software, is largely composed of well-documented, reliable, available routines. This software is readily available to the scientific public, and the journals quickly disseminate research results in numerical algorithms.

If the production code is to have a long life with a minimum of changes, the applications programmer should use the most stable high-quality standardized routines possible. Also, once such a routine is in the local mathematics support library and the production codes use it, many of the difficulties encountered in earlier attempts to solve a similar problem can be avoided. The availability of such routines reduces development costs and can free the applications programmer to spend more time in understanding the areas where uncertainties more specialized to his own problem still exist.

The contents of the library will evolve as research and development permit, with minimal inconvenience to the applications programmer. Trained librarians and system and mathematical programmers provide most of the maintenance so that computer hardware and system changes are usually easy for the user. The user of a good support library can have confidence in a routine and need not know the details of the algorithm. Usually, there is little chance of the ripple effect clobbering the code because good libraries severely restrict the communication links between the routines. This encapsulation of library routines also makes it easy to overlay a large code on a computer that is too small to hold the whole program at once.

One of the keys to writing clear, simple, reliable, and efficient codes is to use the appropriate language. Most applications codes will continue to be written in FORTRAN; however, there is no reason why the library routines need be. True, they must have a FORTRAN subroutine interface, but many of their algorithms rely heavily on constructs that are awkward in FORTRAN. These constructs include logic (such as recursive algorithms), data structures (such as linked lists or unordered sets or stacks), and data types (such as algebraic or multiple precision variables); or they make use of machine architecture intricacies, requiring constructs that allow direct access to the machine hardware features (for example, to keep the pipeline full, to overlap Input/Output (I/O), or to maximize vector chaining). When this is the case, then FORTRAN may be an inefficient and awkward language for the library routines themselves.

Good library routines will help to make efficient use of the hardware without inefficiently using the application programmer's time. They will allow the applications programmers to program in the language they prefer as long as it is compatible with the interfaces provided by the library routines.

However, there are major problems with computer software libraries as they exist today. First, each library has its own quality assurance procedures that a routine must pass before it is accepted. These standards differ from computer center to computer center, and often the user is not aware of what code verification a particular routine has undergone. The computer center librarians rarely make the code assessment document available, causing knowledgeable users to view the software with suspicion.

A second major problem in math software that comes from different sources is the lack of unified documentation, data structures, naming conventions, and coding styles. The current proliferation of styles could scarcely be more confusing if the computer libraries had been organized in the Tower of Babel. Even though each routine may be well documented and reliable and have a stable user interface, the library routines rarely have the same data structure as the applications code. This situation forces the users to continually write interface routines to translate their data structures to those of the library routines.

The third major defect in current computer software libraries is the lack of high-level routines to support physics production codes. The libraries are rich in low-level math routines (such as trigonometric functions) and have a moderate number of middle-level routines (such as those solving linear systems of equations or integrating ordinary differential equations), but there is almost no high-level software.

Such high-level routines would perform many of the common procedures done by most physics codes. For example, the physics models that involve, in an essential way, the numerical solution of complicated systems of partial differential equations[3-5] can be divided into sections: grid generation; grid rezoning or adaptation to the solution or domain boundary; numerical approximation of derivatives; incorporation of boundary conditions; integration of the large, sparse systems equations as they evolve in time; and the solution of large, sparse nonlinear systems of algebraic equations. Most computer libraries have little or no support software in these areas.

One reason such software does not exist is the extra cost and time involved in making a routine general enough to be in a library. Most large-scale computer efforts are funded to solve a specific problem as quickly and as efficiently as possible. The manager in charge of that project is not expected to jeopardize its short term success for some future goals. These managers do recognize that software is expensive and should be developed only once, but the expense to develop it again will be borne by a future project and is, therefore, less relevant.

If a library of high-level support routines did exist, new routines would be immediately accessible to the applications programmers as soon as they were added to the library. As the library expanded, more parts of new codes could be handled by library routines. As a result, the remaining portion that the applications programmer works with will become simpler and more versatile; eventually he will be able to concentrate more on organizing and managing the routines in the library to solve his specific problem than on programming new routines outside his areas of expertise.

With a well-designed applications code, the user can try different combinations of numerical and physical models in the library, tailor the code and optimize it to a specific problem. Also, because the source codes of the library routines are available, users can easily modify these codes to develop and test new ideas on the library code validation test problems, as well as on their own full-scale production problems. This approach also allows specialists in physics, numerical analysis, or computer science to explore new models, methods, and systems software in production codes without being required to know the details lying outside their specialty. Historically, scientists have made many innovations when tinkering in areas outside their specialty; this approach promotes crossfertilization.

Another advantage of the high-level library concept is that it lends itself to small-scale testing. We can experiment with the design of the different minimodules by writing relatively small programs that show the way toward a more reliable composition of the larger pieces. When combining these prototype library routines, we may be able to identify inconsistencies in the structure early in the process and develop interesting sidelines.

This technique has been tested at Los Alamos in a design code of moderate size and found to be better than were any of the other approaches tried. First, we developed a standard set of data structure and naming conventions that were appropriate for the large system of partial differential equations to be solved. Next, we wrote interfaces to many of the standard library math and graphics routines, translating this data structure into the one required by support routines. A preliminary production fluid dynamics code was then written that incorporated these interfaces. Next, high-level routines were written one by one and put in a FORTRAN library called PDELIB. As each routine became available, we removed the corresponding code from the production code, making it more compact, easier to understand, and usually more efficient. Details of some of the modules in PDELIB are described in Appendix B.

V. FUTURE SOFTWARE DEVELOPMENTS.

Writing high-quality software is not a simple extension of do-it-yourself programming. Scientific computing is a research field just as physics, mathematics, and computer science are. Fundamental advances in the capability and efficiency of future production code software will require research and innovations in numerical methods and software design before the software will perform as advertized over a broad range of input data, compilers, operating systems, and machine hardware. There must be checks in the software to detect and recover from (or report) anomalous situations. Attaining the quality and high standards proposed in Appendix A requires detailed analysis, planning, extensive testing, and comprehensive documenting. Though demanding and costly, these tasks are justified when we consider their influence on future production codes.

We can no longer allow software production to remain a hit-or-miss affair based largely on cottage industries with no unified validation, documentation, or data structures. One solution is to form a national effort with cooperation from universities, industry and the national laboratories to ascertain what software is needed and whether it exists and to implement the routines that are needed most. This effort will necessarily have a programmatic emphasis, but it must also have adequate the funding to support the basic research in numerical analysis and computer science needed to answer questions that arise in designing and implementing library routines. In scientific computing, the emphasis in research is different from that in traditional mathematics; for many of the required routines, the problem is not to prove that a method can work in principle but to find one that produces a reasonable answer quickly, reliably, and efficiently.

The coordinating group members must know the strengths and weaknesses of different numerical approaches; selection of routines should not be based on such reasons as "it was not invented here" or "I'm a finite difference not a finite element man." Knowledge, experience, and analysis rather than opinion and tradition are necessary to the design of a system that can effectively cope with the unexpected. The group must include seasoned numerical analysts.

Progress in numerical methods development and physics modeling cannot be divorced from the implementation of software. This fact often requires rethinking an approach so that it is compatible with current languages and available data structures and can be written in a modularly usable fashion. Therefore, talented software engineers must be in the group.

For their library routines to be compatible with production software, this library development team must have good rapport with the user community. The design decisions made in forming the library will be based on the sensitivity of its development team for the needs of future users. Production code programmers comprise a unique community scattered over isolated computer establishments across the country. My own experiences at Los Alamos suggest that we cannot expect library software to be developed at great distances from production code teams.

Software vendors cannot spend the time, effort, and money to provide high-level support packages for such a small user community. These vendors, and most universities, are too far removed from production code designs to interact with them and obtain the feedback essential to map out a good design. Also, production physics codes are often run on the most powerful computers available at any given time, and for the support library to maximize efficiency, those writing them must have ready access to the machines they will run on.

In general, this requires that the high-level support software should be developed at the same computing centers and tested on the same machines where the physics production codes are written.

A. Mathematics Software Group

The natural question to ask is who is currently writing such software? At Los Alamos we have a large graphics group (approximately 18 staff members) organized in the way I have described. This group has had remarkable success in making easily used state-of-the-art graphics available to production code users. We have no similar effort for numerical software now; in fact, numerical software development is currently done by self-motivated individuals and is left almost completely to chance.

Historically, the national laboratories have been aggressively progressive in acquiring the latest computer hardware but, in the recent past, highly conservative in supporting library-quality mathematics software development. To be effective in the long run, I urge the administrators of these laboratories to use more of their work force to develop quality high-level routines that can be used in many design codes rather than to write specific routines that can be used by only one code. Over the years the former approach has received weaker support than has the latter. The resulting chaos has significantly impeded the construction and effectiveness of our large-scale application codes.

Recently, at Los Alamos we have acquired software from vendors and from SLATEC. The SLATEC math library is a poorly funded, interlaboratory, unified effort to acquire and develop software. Under the SLATEC system, our math library is in much better shape than it was four years ago. During this time, however, the Laboratory has not funded a single new mathematics routine through to implementation. The SLATEC approach gives us some direction, but progress will continue to be slow without adequate funding for developing new software.

If software groups were formed and coordinated in a national effort to develop algorithms and software in a structured way and made responsible for advanced development and research in numerical methods, software design, and implementation of high-level support software, they would soon play a vital role in the development of our future production codes. The teams would have to be adequately staffed to provide timely responses to the legitimate, critical needs of their particular user community. The group should emphasize, recognize, produce, and support high quality software for large-scale scientific computing; they should be programmatic groups with an internal research effort and with a consulting section. Also, this would be an effort not to replace application programmers but rather to support them with auxiliary software tools. These new groups would be less concerned with writing the great American codes; they would be more concerned with constructing the building blocks and tools that will be needed to write them.

Within these groups there should be an advanced development team to guide the project, a team maintaining a basic understanding of relevant proven advances. The team leader should encourage the gathering of basic knowledge in areas of critical need as opposed to areas where any practical application is doubtful. But decisions here must be carefully considered: a short-sighted search for immediate practicality can destroy an individual's initiative to find new, unexpected advances.

Users and application code developers should continually review and advise the software development group. Therefore, the group should publicize their work objectives through progress reports and presentations. Feedback during the development process and constructive criticism are welcome in a healthy development environment; they help prevent costly mistakes. Numerical analysts and software engineers can develop better algorithms and software by collaborating with the applications programmers; a section of the groups should work closely with applications programmers to test the new software in current production codes.

Although the groups should maintain close acquaintance with a variety of applications programmers, they should be separated from the managers of specific applications programs. By divorcing the group from any particular applications effort and defining a broad set of goals, can pursue fundamental improvements in general computing methods as well as in software supporting specific applications. The software produced will be general rather than specific; therefore, its effect will be seen in the many different applications for it.

The new software must receive strong support in the local computer libraries and obtain certification through a national code assessment program. In many organizations this means that the development group would be under the same management as the computer center or under other management not concerned with completing one of the organization's specific missions.

The high-level support library will raise the productivity and precision of future production codes to a level that will appear incredible when compared with the levels encountered in earlier experiences. Equally important will be the positive psychological effect when applications programmers become aware of their enhanced power to write complicated programs correctly and easily using high-level library routines.

B. Laboratory/University Interactions

Knowledge of the behavior of numerical approaches is usually not centralized. The software development group must communicate with future users and the numerical algorithm research community in the national laboratories, in industrial research centers, and at universities. In the last decade, the Federal government has continually reduced its support of large-scale computing by private corporations and at universities. In fact, to my knowledge, no US university had installed a state-of-the-art scientific supercomputer of the early 1970s (a CDC 7600); only two universities now have easy access to a current scientific supercomputer (CRAY-1 or CDC 205). Universitiy scientists and students need to develop a sense of what scientific computing is like today, not what it was in bygone days of expensive, memory-limited, slow hardware. The immense computing power available at the national laboratories is usually unavailable in industry and universities. Indeed, the most sophisticated scientific computing centers in the US exist only in the national laboratories, such as Los Alamos, Lawrence Livermore, the Livermore Magnetic Fusion Energy Center, and the National Center for Atmospheric Research. Interactions between production code developers and university researchers are most useful when the university scientists and their students have access to the state-of-the-art computers on which the codes will be run. Today, such access is virtually nonexistent.

Without this access, even though they may desire to, university scientists are often unable to research new physics models and numerical algorithms to develop polished software optimized on scientific supercomputers. The granting agencies recognize this, and today they rarely require such researchers to produce quality software that can be used by others.

Applications programmers at the national computer centers sometimes complain that university computer scientists propose solutions to largely irrelevant problems. Because access to large-scale computers is lacking, these university scientists work in a smaller environment and tend to develop new models, methods, and systems that are appropriate for smaller problems. The algorithms in large-scale scientific codes can be highly complicated and problem dependent. That is, a method developed for a particular test problem may not work for similar but larger problems. Methods that work well in one space dimension may not be easily extended to two or three dimensions. Linear analysis can rarely ensure accuracy with nonlinear methods or highly nonlinear equations. This fact puts a huge burden on production code developers, who must review the scientific literature, full of small-scale test problems solved by simple methods, to choose algorithms that will scale properly to model a far more complex system. After the choice is made, they must redevelop the software necessary to test it—usually duplicating the efforts of the original author.

This trend must stop if we are to maintain the excellence needed to lead in future scientific computer modeling.

We must vigorously and aggressively initiate a long-term, large-scale scientific computation program that encourages multidisciplinary interactions between the universities and the national computing centers. The national laboratories must provide easy long-distance access to their supercomputers, encourage active visiting scientist and graduate student programs, and sponsor workshops and conferences on the currently important scientific questions in large-scale computing. Many misunderstandings could turn into collaborations if university scientists and application programmers worked together.

VI. CONCLUSIONS.

Large-scale scientific calculations that tax the resources of the most powerful computers will continue to be essential to modern research and development efforts. To obtain long-term reliability and stability of future physics production codes, we must coordinate designing, implementing, and testing of high-level numerical software. We must set priorities and goals and support a strong research and development effort with cooperation among applications programmers, the theoretical numerical methods researchers, and the computer science software developers.

The complexities of modern programming force us to shift from the superprogrammer teams of the past to software engineering teams. This shift will require strong support from management and reorientation of any programmers who are set in the traditional ways. Programmers, no matter their education, skill, or bias, need to accept the importance of better planning and the need to specialize for the sake of creating a more efficient division of labor.

In the early days of aeronautics, to build an airplane you needed a superstar, like one of the Wright brothers. Gradually, the procedure became well known, and today, any handbook engineer can build an airplane in his garage with a mail-order kit. However, to build a space shuttle, we still need superstars. In the early days of scientific computing, getting a multidimensional aeronautics code running required a superprogrammer. But today many of these techniques are well known, and our superprogrammers should be free to work on the equivalent of the space shuttle. Unless this is done, a crisis in production code development will soon be here. Already there are too few superprogrammers to write the next generation of physics production codes. Many large corporations and government laboratories have a backlog of to-be-developed programs, partly because many of today's best programmers spend their time maintaining old programs and duplicating the efforts of other superprogrammers on similar projects.

We must avoid repetition of expensive, error-prone, and time-consuming coding of commonly used methods. Much of what is now done is redundant and could be eliminated if a library of high-level subroutine support packages were available. If the common elements of the code are available as modules, the applications programmers can plug into these routines and eliminate much of their effort. The library routines have been carefully tested, and as they are used, more and more of the errors will be weeded out, making the modules even more reliable.

With any new project, we must weigh the importance of the project and its potential for success against the costs. Supporting a national effort to fund and coordinate active programmatic groups to develop high-level numerical software creates a high-leverage, cost-effective situation with a guaranteed payoff of software spinoffs to the scientific community in industry and universities. Saved computer time, shorter production code development time, more effective use of applications programmers, visible positive spinoffs to industry, and more reliable production designs will repay the initial investment many times.

The proposed guidelines are based on simple, tried-and-true principles found to reduce errors in programs, to reduce the costs and development time of future codes, and to increase future code capabilities. Structured code design and testing, using a high-level support library, have an excellent chance of resulting in an applications code that performs as expected from the beginning of its production lifetime. The development of software tools that support production physics codes will be a major basis for technological improvements. We have the unique opportunity to design the tools and define the standards that will form the basis for the next generation of codes; it is unique because, once a standard has been set, it will be as hard to change as going from English to metric units in the middle of building an automobile. Users develop mechanisms to compensate for the deficiences in any system. Once established, they are hard to change.

The present direction of our national scientific production-code community will not lead to an acceptable future. We propose this effort so that new developments can be made available to applications programs quickly—avoiding the currently long waiting time required for them to filter down by chance. If we fail, taking full advantage of breakthroughs in computer hardware and numerical algorithms in a timely fashion will be impossible. There is a sense of frustration in seeing the years running quickly by without our future prospects improving. The obstacles are no greater now than they were years ago, but there is less time to surmount them: international competition has upped the stakes. Today, time is more a threat than a promise.

ACKNOWLEDGMENTS

I sincerely thank Blair Swartz, Bob Boland, Dan Carroll, Stirling Colgate, Joel Dendy, Paul Frederickson, Debby Hyman, Darryl Holm, Tom Manteuffel, Paul Stein, Burton Wendroff, and Andy White for their comments and advice.

REFERENCES

1. B. Buzbee, "Japanese Supercomputer Technology," Los Alamos National Laboratory document LA-UR-82-2733, submitted to Science, American Association for the Advancement of Science, Washington, DC (1982).
2. P. D. Lax Committee, "Draft Report of the Panel on Large Scale Computing in Science and Engineering" (Code 9/29 Version 5), under the sponsorship of the Department of Defense and the National Science Foundation in cooperation with the Department of Energy and the National Aeronautics and Space Administration (September 30, 1982).
3. Committee on the Applications of Mathematics, "Computational Modeling and Mathematics Applied to the Physical Sciences," draft, Office of Mathematical Sciences, Commission on Physical Sciences, Mathematics, and Resources, National Research Council, National Academy Press (October 25, 1982).
4. Hans G. Kaper, "Program Directions for Applied Analysis," Applied Mathematics Division, Argonne National Laboratory, Argonne, Illinois (January 1980).
5. "Program Directions for Computational Mathematics," Applied Mathematical Sciences, Robert E. Huddleston (Chairman), Division of Engineering, Mathematics, and Geo-Sciences, Office of Basic Energy Research, US Department of Energy (1980).

The author was supported by the US Department of Energy, Division of Basic and Engineering Sciences under contract KC-04-02-01 and the United States Department of Energy under contract W-7405-ENG-36.

Center for Nonlinear Studies
Theoretical Division, MS B284
Los Alamos National Laboratory
Los Alamos, NM 87545

APPENDIX A
QUALITY ASSURANCE STANDARDS

I. INTRODUCTION.

Programmers make errors. When writing a code to be used by others, programmers should protect users from inadvertent programming errors by a code certification plan. A good code certification plan usually has few restrictions on the conventions of the innermost syntax, a sensitive subject with most programmers. (It is here that personal style manifests itself.) Instead, the plan consists of reasonable standards that improve the reliability and documentation of a code. If the code is to be part of a larger system, the plan may dictate the allowed data structures, the allowed connections between the module and the user's code, and perhaps the style of these connections, all of which enhance the overall consistency of the system.

II. RELIABILITY.

Most programming time in a physics production code goes into debugging and maintenance—not physics. To reduce these costs, any supporting module must be as error-free as possible and undergo stringent test procedures, based on sound, practical technical standards rather than on good ideas to be used when convenient. A code is reliable to the extent that our experience makes us confident in it, and as time goes on, we find no bugs. Demonstrating correctness by sample testing will show only the presence of errors, not their absence. Sound testing methods, however, are still our best reliability check and must be used after every module change.

Using standard error handlers with low-cost internal consistency and argument checks, counters at key points, and global debugging variables can help find programming errors. During the design phase, a test routine should be written to check the efficiency and correctness of the results in a large class of realistic problems with known characteristics. This test set must be realistic because the results may not generalize outside their immediate range. Users should have ready access to the test routine and code certification document to compare the effectiveness of existing algorithms with new ones.

The validation procedures go beyond a code certification team testing the modules that compose the production code. The applications programmer must rigorously test the interconnections between the modules. Often, production codes have additional validity checks based on conservation laws, convergence tests, and different formulations of the physical model or the numerics. These tests are usually based on global properties of the model and cannot be performed by the module alone. They cost little in computer time and effort compared with the increased reliability they give the applications code.

Although everyone agrees that reliability is more important than execution time, machines will continue to be finite for physics production codes, and we cannot ignore how fast the code executes.

III. EFFICIENCY.

Most production codes spend a large part of the execution time in a small part of the code. The ideal approach is to first construct a correct code, then locate those program parts that are most heavily used and optimize them to improve the overall performance. If the code is modular, the performance can be compared easily by switching formulations in the most used module. Because these modules can be optimized and debugged separately, the chances of introducing new errors in the production code are reduced when these different modules are integrated into the main code. If the main code is not modular, the programmer often sacrifices hardware efficiency because of lack of time or expertise. When modules are used, we can let the experts optimize the coding, using algorithms that are highly dependent on the computer hardware architecture.

For this approach to work, the experts writing the modules must have easy access to the supporting software and the machine architecture. The language they use need not be the same as the applications code because they will need to be able to push the hardware to its potential. Parallel processing, vector chaining, memory conflicts, and storage registers will continue to be important, and the use of modules keeps such things from being foremost in the applications programmer's mind.

However, applications programmers must be careful not to hide knowledge of parallelism or synchronization from the modules. Programmers should use their knowledge to make it easier for the module to execute the most efficient code. A flag may be needed to allow the algorithm used in the module to respond to the machine architecture whenever possible.

Types of computers and operating systems to run the applications codes will proliferate in the next few years. We must anticipate these changes and design the interface between the module and applications code to be as stable as possible. In this way the applications code can be easily moved from machine to machine, always using the modules that are best suited to and optimized for the current computer hardware. It is a major task to design support software with a stable interface and still fully exploit the available machine hardware. One criterion is clear though—the data flow between the module and the applications code must be severely restricted.

IV. DATA FLOW.

Every connection between modules is a path along which errors can propagate. The connections between the module and the outside world in FORTRAN should be limited to explicit subroutine-calling-sequence arguments whenever possible. Because of computer language limitations or for efficiency, within the module it may be necessary to use common blocks local to the module. These common blocks may also contain useful expository and diagnostic information that allows the user to increase a routine's efficiency.

V. DOCUMENTATION.

Clear documentation is most important if a code is to be used correctly the first time. No matter how good a routine is, it will never be used to its full potential without good documentation, even when the new software is much better than other routines in use.

For a large project such as a physics production code, the supporting modules should have consistent documentation, user interfaces, and variable names. By planning for future changes, the support staff can correct errors, adapt the system to hardware and software changes, enhance capabilities of the routines, and improve the internal structure while preserving the external interface. When routines are constantly being improved, they continue to look like but work better than the old ones.

The user's manual should be written during code design because it helps everyone agree on the functions to be performed by the system and the user. It must fully specify the routine's function and the algorithm as well as the input, output, and error-flag variables. Later, in the construction phase, the manual is a reference necessary to verify the code.

The documentation should have many levels. For the casual user, reading only a few pages should be enough for him to easily access relatively sophisticated features. On the other hand, the sophisticated user needs a detailed code description and the capability to tune the routines and optimize them for his specific problem.

The documentation must give useful information on the efficiency, portability, and appropriateness of the different algorithm combinations for various classes of problems. The efficiency for small problems should be included but in many cases is irrelevant. Because the performance of the routines depends so heavily on the problem size and implementation and on the computer architecture, only users who have tried the routines on large production problems can report on efficiency and effectiveness. There should be a way to update the documentation with such reports as they become available. Exhaustive testing of all the options on large problems is usually impractical because of time and cost limitations.

All code documents should have a complete methods section so that a competent programmer can rewrite the code.

Many (probably most) readers of the code will skip this section because they know the methods used or have no interest in the implementation detail. A cornerstone of the scientific method is that the results from good science must be reproducible, and careful writing of this section is critically important if the code is to represent good science. During a quality assurance review, a good reviewer will read the methods section carefully. If he cannot assure himself that he could repeat the coding, he should recommend rejection of the code—no matter how awe-inspiring the results.

APPENDIX B
HIGH-LEVEL MATHEMATICS LIBRARY

I. INTRODUCTION.

PDELIB, an early attempt to develop a high-level mathematics library at Los Alamos, contains both general, robust methods and methods developed over the years that are highly effective on a specific set of partial differential equations. The various solution algorithms for large systems of partial differential equations were first simplified so we could systematically study the general patterns of interconnections and the flow of the algorithms. Then, methods were unified under a single theory and made available in a coherent library. Based on this experience we now describe some major modules that might be available in a high-level mathematics library of the future.

II. GRID GENERATION.

The input for the grid generation routines is the grid type [for example, a tensor product grid (x_i, y_j), a multiple argument grid $(x_{i,j}, y_{i,j})$, or a neighborhood grid such as a triangular or tetrahedral grid], the domain boundaries, and any additional optional information, such as the grid distribution along designated lines or regions. The user can also specify parameters that control the smoothness of the grid variations. The grid generation routines then generate a boundary-fitted grid with the requested number of points in the domain. The routines can be used to generate a grid along a surface, such as the boundary of a region; and later, this boundary grid can be used as a reference surface when the interior grid points are defined.

III. REZONING.

In the rezoning routines, the input is a grid and a set of functions defined at the grid points. The routine generates a new grid that better resolves the functions either with a fixed number of grid points or within a specified accuracy with as few points as possible. The user can specify a performance index function that estimates the goodness of the original grid. The new grid would then be chosen so that the performance index is equally good everywhere. The new grid can be made to agree with the original grid at certain prescribed reference grid points, such as those along boundaries.

IV. INTERPOLATION AND DIFFERENTIATION.

The input for these routines includes a function defined at a discrete set of grid points along with another set of points when an approximation to the function or its derivatives is advisable. The code then uses an interpolant to provide these results, along with error estimates (if requested).

The second set of grid points might be the output from the rezoning routine; or if the original data was a table, the second set could be the table lookup points. When a smoothing interpolant is used, such as local least squares, then these routines provide a simple way to filter data. The interpolation is also highly useful in graphics routines.

Adaptive mesh methods require accurate, smooth multi-dimensional interpolation methods on nonuniform grids. These methods are difficult to program in two and three dimensions, and in many instances the theory is weak and no available, reliable algorithms exist. Clearly, whatever algorithms emerge in the future will not be simple, and it is unrealistic to expect applications programmers to devote the time and energy necessary to lead the research and software effort in this highly specialized field of approximation theory.

The high-level library will allow the approximation theorists to test their new methods in realistic situations.

Many of the available supporting input and output routines (such as graphics routines) require special data structures that are not used by the physics code. Instead of having to restructure the physics codes to suit the graphics package, we use these interpolation routines that make it easy to write an interface routine that switches back and forth between different data structures. Also, these routines allow users to switch between one-, two-, or three-dimensional data structures easily.

Numerical differentiation is a critical element in any code based on partial differential equations; and these supporting routines provide stable, accurate, and reliable approximations to derivatives, using local polynomial interpolation methods, finite element methods, or global pseudospectral methods. These routines also help the user handle many of the delicate mesh-boundary-interface troubles that are so overwhelming in three-dimensional calculations.

V. BOUNDARY CONDITIONS.

Computing accurate boundary conditions for a system of partial differential equations is as important as accurately approximating the spatial derivatives in the interior of the domain. Boundary conditions exert one of the strongest influences on the behavior of the solution. Also, the errors introduced into the calculation from improper boundary conditions persist even as the mesh spacing tends to zero. If the boundary conditions are not properly incorporated into the discrete approximation, a well-posed problem in differential equations can be changed into an ill-posed (unstable) discrete one.

These routines use the well-established techniques based on extrapolating the numerical solution from inside the domain to fictitious points outside the domain or at the boundary. The extrapolant is constructed so that the resulting discrete approximation at the boundary is consistent with as many relationships that can be derived from the boundary conditions and differential equation as possible.

The user supplies the solution inside the domain, a differential relationship that must be satisfied at the boundary, and the type of approximation to be used. For hyperbolic systems the routine might also ascertain if the boundary conditions are well posed.

Modeling complex curved boundaries in two and three dimensions is a challenge—the extrapolation formulas are difficult to derive and tedious and error prone to program. At best, most current two-dimensional codes have a fair-to-poor approximation when the domain has sharp corners and material interfaces; in three dimensions it is not even clear what should be done. Necessarily, the software for these complicated boundaries will be tied closely to a strong research effort.

VI. TIME INTEGRATION ROUTINES.

The numerical solution of evolutionary differential equations is advanced in time in discrete steps that vary, depending on the local behavior of the solution; that is, the length of the time steps depends on whether the solution is evolving slowly or quickly. The methods that approximate the time derivatives, like those that approximate the space derivatives, are based on Taylor series. The major difference between time and space differentiation is that time has a direction. This time flow allows savings in computer storage but introduces questions about time stability in the difference equations relative to the stability of the differential equation.

The routines in this module require the user to provide a procedure to define the time derivatives of the ordinary or partial differential equations as a function of spatial derivatives. The routine then advances the equations in time, based on user-supplied accuracy or stability criteria. The user has full control over the methods used and the degree of implicitness with which the equations are solved, or the routines will adaptively select a method, depending on the equations being solved.

Implicit methods require solving one or more nonlinear algebraic systems on each trip. This can be done automatically, or the system can be constructed and the responsibility for computing accurate solutions left to the user.

VII. ALGEBRAIC EQUATION SOLVERS.

The numerical solution of large, sparse nonlinear algebraic systems, arising in implicit approximations to partial differential equations, is often obtained by function iteration. The iteration may be based on a simple defect-correction method such as successive overrelaxation (SOR), the alternating direction implicit method, an incomplete (LU) factorization, or a multigrid method. The routines in this module have software to solve large-scale linear problems automatically, using one of these methods as specified by the user.

To accelerate convergence of an iterative algorithm, we can use a method such as Chebyshev or a conjugate gradient algorithm. For example, if the applications code were based on iteration of a SOR-like method and if the iteration acceleration routine were called after each cycle, we could improve the iteration to a SOR-Chebyshev method and converge much faster. This would require only a one line change in the applications code—the subroutine call.

VIII. INPUT/OUTPUT ROUTINES.

The main advantage of these routines is that they all have the same data conventions. They speak the same language to the user or the other modules and then translate it into whatever is needed for the graphics routines, such as Display, BYU-Movie, or the NCAR package. Also, there are many printer routines with clear and simple output formats for printing the solution in one, two, and three dimensions on uniform and nonuniform grids.

The heart of the I/O routines is a standardized metafile and setup utility that can be used to communicate between different codes and link them together. The setup utility contains a highly interactive graphics and rezone program that can be used to initialize the problem or change the data structure (such as going from two to three dimensions). The setup utility makes much use of the grid generation, rezone, and interpolation modules. The metafile dumps can be used for graphics postprocessing or restarting the problem at any dropfile time, using the setup utility to make any desired changes.

Los Alamos
Los Alamos National Laboratory
Los Alamos, New Mexico 87545

RECENT MATHEMATICAL AND COMPUTATIONAL DEVELOPMENTS IN NUMERICAL WEATHER PREDICTION

Akira Kasahara

1. INTRODUCTION

Meteorology is an old science, but the techniques of weather prediction have only been revolutionized since the advent of high-speed computers [110, 94]. The art of weather prediction is now a large-scale computing problem based on hydrodynamic principles. Recently, there have been significant improvements in forecasting large-scale mid-tropospheric flow patterns in the short- to medium-range period, up to 7 days or so [11]. These may be partly due to improved atmospheric observations and analyses resulting from the Global Weather Experiment [41, 42] conducted in 1979, but our increased theoretical understanding of the planetary-scale motions and its application to operational forecasting has undoubtedly contributed to this cause. In this review, I present recent mathematical and computational developments in numerical weather prediction which, I believe, have contributed to improved large-scale weather forecasting.

2. GLOBAL ATMOSPHERIC PREDICTION MODEL

Motions of the atmosphere are governed by physical laws that may be expressed as the equations of hydrodynamics and thermodynamics. In describing global-scale motions, we use spherical coordinates with longitude λ and latitude ϕ.

ISBN 0-12-546080-5

Because the global-scale motions are quasi-horizontal, their vertical acceleration is negligible and the hydrostatic equilibrium can be assumed, i.e.

$$\partial p/\partial z = -\rho g, \tag{2.1}$$

where p denotes the pressure, ρ the density and g the acceleration due to the earth's gravity. Hence, pressure p may be used as a vertical coordinate instead of height z.

The horizontal equations of motion for the longitudinal and meridional velocity components u and v, respectively, may be written as

$$\frac{du}{dt} - (f + \frac{u}{a}\tan\phi)v = -\frac{g}{a\cos\phi}\frac{\partial z}{\partial\lambda} + F_\lambda, \tag{2.2}$$

$$\frac{dv}{dt} + (f + \frac{u}{a}\tan\phi)u = -\frac{g}{a}\frac{\partial z}{\partial\phi} + F_\phi, \tag{2.3}$$

with

$$\left.\begin{aligned} \frac{d}{dt} &= \frac{\partial}{\partial t} + \mathbf{V}\cdot\nabla + \omega\frac{\partial}{\partial p} \\ \mathbf{V}\cdot\nabla &= \frac{u}{a\cos\phi}\frac{\partial}{\partial\lambda} + \frac{v}{a}\frac{\partial}{\partial\phi} \end{aligned}\right\}, \tag{2.4}$$

where t denotes time, a represents a mean radius of the earth, and $f = 2\Omega\sin\phi$ with Ω being the earth's angular velocity of rotation. The measure of vertical velocity is expressed by $dp/dt \equiv \omega$. In (2.2) and (2.3), F_λ and F_ϕ represent the longitudinal and meridional components, respectively, of frictional force per unit mass.

The mass continuity equation is expressed as

$$\left.\begin{aligned} \nabla\cdot\mathbf{V} + \partial\omega/\partial p &= 0 \\ \nabla\cdot\mathbf{V} &= \frac{1}{a\cos\phi}\left[\frac{\partial u}{\partial\lambda} + \frac{\partial}{\partial\phi}(v\cos\phi)\right] \end{aligned}\right\}. \tag{2.5}$$

The first law of thermodynamics may be expressed in the form

$$d\ln\theta/dt = Q/(C_pT), \tag{2.6}$$

where θ is the potential temperature,

$$\theta \equiv T(p_{oo}/p)^\kappa \tag{2.7}$$

with $p_{oo} = 1013$ mb and T is the temperature given by the ideal gas law

$$p = \rho RT \qquad (2.8)$$

in which R represents the specific gas constant. In (2.7), $\kappa = R/C_p$ where C_p stands for the specific heat at constant pressure and Q denotes the rate of heating/cooling per unit mass. Substitution of (2.7) into (2.6) yields

$$\frac{\partial T}{\partial t} + \mathbf{V} \cdot \nabla T + \frac{\omega T}{\theta} \frac{\partial \theta}{\partial p} = \frac{Q}{C_p}. \qquad (2.9)$$

Lastly, elimination of ρ between (2.1) and (2.8) yields

$$g\ \partial z/\partial p = -\ RT/p. \qquad (2.10)$$

The system of equations (2.2), (2.3), (2.5), (2.9) and (2.10) constitutes the basic set of global prediction equations, governing the time evolution of the dependent variables u, v, T, ω and z. The boundary conditions are that $\omega = T = 0$ at $p = 0$ and $dz/dt = 0$ at pressure p_s near the earth's surface. The physical processes of the atmosphere are contained in the frictional terms F_λ anf F_ϕ and the heating/cooling term Q, which must be expressed as functions of the dependent variables. Although we are not concerned here with the effect of the earth's orography for the sake of brevity, computational methods are available to take into account its important effect in the global prediction models [88, 56].

3. NORMAL MODES

We can learn a great deal about the characteristics of global atmospheric motions by examining the solutions to a linearized system of the basic equations presented in Section 2. A simple perturbation system is the one linearized around the atmosphere at rest with a basic state temperature T_o which depends only on p. Solutions to such a system are referred to as free motions. The characteristics of free motions in the linearized system are called normal modes [67]. In this section, we will discuss the normal modes of the global atmospheric model.

The set of basic equations presented in Section 2 can be written in the form

$$\frac{\partial u}{\partial t} - 2\Omega \sin\phi\, v + \frac{g}{a \cos\phi} \frac{\partial z}{\partial \lambda} = N_u, \tag{3.1}$$

$$\frac{\partial v}{\partial t} + 2\Omega \sin\phi\, u + \frac{g}{a} \frac{\partial z}{\partial \phi} = N_v, \tag{3.2}$$

$$\frac{\partial}{\partial t} \left[\frac{\partial}{\partial p} \left(\frac{g}{S_o} \frac{\partial z}{\partial p}\right)\right] - \nabla \cdot \mathbf{V} = N_z, \tag{3.3}$$

where

$$\left.\begin{aligned} N_u &= -\mathbf{V} \cdot \nabla u - \omega \frac{\partial u}{\partial p} + \frac{uv}{a} \tan\phi + F_\lambda \\ N_v &= -\mathbf{V} \cdot \nabla v - \omega \frac{\partial v}{\partial p} - \frac{u^2}{a} \tan\phi + F_\phi \\ N_z &= -\frac{\partial}{\partial p} \left(\frac{g}{S_o} \mathbf{V} \cdot \nabla \frac{\partial z}{\partial p} + \frac{S'}{S_o} \omega\right) - \frac{\partial}{\partial p} \left(\frac{\kappa Q}{S_o p}\right) \end{aligned}\right\}. \tag{3.4}$$

Eqs. (3.1) and (3.2) are identical to (2.2) and (2.3), respectively. Eq. (3.3) is derived by combining (2.9) and (2.10) and defining

$$S_o = g(dz_o/dp)(d\ell n\theta_o/dp) \tag{3.5}$$

as the mean static stability calculated from the basic temperature profile T_o. Quantity S' denotes the departure of the static stability $g(\partial z/\partial p)(\partial \ell n\theta/\partial p)$ from the mean value S_o.

We now construct the normal modes of the model atmosphere by obtaining the solutions of the linearized system without the rhs of (3.1) - (3.3). The linearized system is

$$\left.\begin{aligned} &\frac{\partial u'}{\partial t} - 2\Omega \sin\phi\, v' + \frac{g}{a \cos\phi} \frac{\partial z'}{\partial \lambda} = 0 \\ &\frac{\partial v'}{\partial t} + 2\Omega \sin\phi\, u' + \frac{g}{a} \frac{\partial z'}{\partial \phi} = 0 \\ &\frac{\partial}{\partial t} \left[\frac{\partial}{\partial p} \left(\frac{g}{S_o} \frac{\partial z'}{\partial p}\right)\right] - \nabla \cdot \mathbf{V'} = 0 \end{aligned}\right\}, \tag{3.6}$$

where the prime over the dependent variables indicates those of the linearized system. The corresponding upper and lower boundary conditions are expressed by

$$\partial z'/\partial p = 0 \qquad \text{at } p = 0 \tag{3.7}$$

and

$$\frac{\partial z'}{\partial p} - \frac{d\,\ell n\,\theta_o}{dp} z' = 0 \qquad \text{at } p = p_s. \tag{3.8}$$

The linearized system (3.6) with boundary conditions (3.7) and (3.8) are solved by the method of separation of variables. We assume that the dependent variables are expressed by

$$\begin{pmatrix} u' \\ v' \\ z' \end{pmatrix} \propto \begin{pmatrix} \tilde{u} \\ \tilde{v} \\ \tilde{z} \end{pmatrix} G(p), \tag{3.9}$$

where G(p) is the vertical structure function common to all the three dependent variables. Variables $\tilde{u}$, $\tilde{v}$, and $\tilde{z}$ are horizontal structure functions of λ, ϕ, and t.

By substituting (3.9) into (3.6) and applying the separation of variables, we obtain

$$\left.\begin{aligned} &\frac{\partial \tilde{u}}{\partial t} - 2\Omega \sin\phi\, \tilde{v} + \frac{g}{a\cos\phi}\frac{\partial \tilde{z}}{\partial \lambda} = 0 \\ &\frac{\partial \tilde{v}}{\partial t} + 2\Omega \sin\phi\, \tilde{u} + \frac{g}{a}\frac{\partial \tilde{z}}{\partial \phi} = 0 \\ &\frac{\partial \tilde{z}}{\partial t} + D\, \nabla \cdot \tilde{\mathbf{V}} = 0 \end{aligned}\right\}, \tag{3.10}$$

where the separation constant D satisfies the ordinary differential equation

$$\frac{d}{dp}\left(\frac{g}{S_o}\frac{dG}{dp}\right) + \frac{G}{D} = 0, \tag{3.11}$$

which is referred to as the <u>vertical structure equation</u>. The boundary conditions (3.7) and (3.8) are expressed as

$$dG/dp = 0 \qquad \text{at } p = 0 \tag{3.12}$$

and

$$\frac{dG}{dp} - \frac{d\,\ell n\,\theta_o}{dp} G = 0 \qquad \text{at } p = p_s. \tag{3.13}$$

System (3.10) presents the <u>horizontal structure equations</u> and has the same form of the linearized shallow-water equations over the sphere for a mean fluid depth D. Extending the work of Laplace [20] and Lamb [67], Taylor [107] treated the oscillations of an atmosphere over the rotating earth and showed that there is an infinite set of values for the separation constant D. The parameter D is referred to as <u>equivalent height</u>.

The value of D in system (3.11) is determined as the eigenvalue of (3.11) under boundary conditions (3.12) and (3.13), constituting a Sturm-Liouville problem. Vertical structure function G(p) is the eigenfunction of (3.11) associated with an eigenvalue D_n. It can be shown that the normalized eigenfunctions G_i and G_j, corresponding to eigenvalues D_i and D_j respectively, when $D_i \neq D_j$, are orthogonal in the sense that

$$p_s^{-1} \int_0^{p_s} G_i(p)\, G_j(p)\, dp = \delta_{ij} , \tag{3.14}$$

where $\delta_{ij} = 1$ if i = j and zero otherwise. The values of D and the corresponding eigenfunctions G(p) for an atmospheric model similar to the one presented here are given in [63]. The largest value of D for the standard atmosphere is approximately 10 km and the corresponding vertical structure function G(p) varies little with respect to p. This gravest mode is called the external mode. The remaining modes, called internal modes have a smaller value of D and the corresponding eigenfunctions G(p) show increased variations with pressure for decreasing values of D. The external mode is the only discrete physically significant free mode in an atmosphere with realistic distributions of the earth's atmospheric temperature and dissipation [69]. Internal modes appear as a continuous spectrum. For the internal modes, the values of D and the corresponding eigenfunctions vary depending on the number of internal nodes to be obtained.

The horizontal structure equations (3.10) are identical to Laplace's tidal equations [20] without tide-generating functions. We introduce new dimensionless variables

$$\hat{u} = \tilde{u}/(gD)^{1/2}, \quad \hat{v} = \tilde{v}/(gD)^{1/2}, \quad \hat{z} = \tilde{z}/D,$$
$$\hat{t} = 2\Omega t, \tag{3.15}$$

and rewrite (3.10) in the form

$$\partial \hat{\mathbf{W}} / \partial \hat{t} + L\hat{\mathbf{W}} = 0, \tag{3.16}$$

where $\hat{\mathbf{W}}$ denotes the vector dependent variable

$$\hat{\mathbf{W}} = (\hat{u}, \hat{v}, \hat{z})^T$$

and L is the linear differential matrix operator

$$\mathbf{L} = \begin{bmatrix} 0 & -\sin\phi & \frac{\gamma}{\cos\phi}\frac{\partial}{\partial\lambda} \\ \sin\phi & 0 & \gamma\frac{\partial}{\partial\phi} \\ \frac{\gamma}{\cos\phi}\frac{\partial}{\partial\lambda} & \frac{\gamma}{\cos\phi}\frac{\partial}{\partial\phi}[\cos\phi(\,)] & 0 \end{bmatrix} \tag{3.17}$$

in which

$$\gamma = (gD)^{1/2}/(2\Omega a) \quad (\equiv \varepsilon^{-1/2}) \tag{3.18}$$

is a dimensionless parameter characterizing the nature of shallow-water flows. A related quantity $\varepsilon = \gamma^{-2}$ is called Lamb's parameter [70].

The solution $\hat{\mathbf{W}}$ of Eq. (3.16) can be expressed as a linear combination of functions that have the form

$$\hat{\mathbf{W}}(\lambda,\phi,t) = \mathbf{H}(\lambda,\phi)\exp(-i\,\hat{\sigma}\,\hat{t}), \tag{3.19}$$

where $H(\lambda,\phi)$ represents the horizontal structure function of normal mode and $\hat{\sigma}$ is the corresponding dimensionless frequency scaled by 2Ω. Substituting (3.19) into (3.16), we see that $H(\lambda,\phi)$ is an eigenfunction of L and $\hat{\sigma}$ is the corresponding eigenvalue of $\mathbf{L}$, i.e.

$$\mathbf{L}\,\mathbf{H} = i\,\hat{\sigma}\,\mathbf{H}\,. \tag{3.20}$$

Because the coefficients of $\partial/\partial\lambda$ in (3.17) are independent of λ, the function $H(\lambda,\phi)$ is expressed by

$$\mathbf{H}(\lambda,\phi) = \circledH(\phi)\,e^{is\lambda}, \tag{3.21}$$

where s denotes the zonal wavenumber and $\circledH(\phi)$ is the meridional modal function which depends only on latitude ϕ. There are a number of meridional modal functions corresponding to each zonal wavenumber s and we use index ℓ to denote the ℓth meridional normal mode. Hence, the horizontal structure of a normal mode $\mathbf{H}(\lambda,\phi)$ and the associated frequency $\hat{\sigma}$ depend on the zonal wavenumber s and meridional index ℓ in addition to Lamb's parameter ε.

The history of obtaining the solutions to Laplace's tidal equations (3.10) is of some interest. Laplace [20] first treated oceanic and atmospheric tidal problems and formulated the linearized shallow-water equations (3.10) which now bear his name. Laplace obtained their solutions by expansion in a power series of trigonometric functions. Margules [80] also used the same approach, but investigated solutions to free oscillations. A few years later, apparently unaware of Margules' work, Hough [53] employed a series of associated Legendre functions to calculate the normal modes of (3.10) and obtained a faster rate of convergence of solutions than the use of trigonometric functions. After the pioneering work of Margules and Hough, solution of Laplace's tidal equations remained as a challenging mathematical physics problem [67] and was treated mostly in connection with atmospheric tides [26].

Despite the importance of the problem in application to the atmosphere and ocean, the full scope of the properties of the normal modes had not been explored until the 1960's, due partly to the need for extensive numerical calculations which became possible after the advent of electronic computers. Extensive investigations appeared during the 1960's (see the references in [70]). In particular, Longuet-Higgins [70] published tables of eigenfrequencies and diagrams of eigenfunctions for a full range of ε together with discussions on the asymptotic properties of the solutions for limiting values of ε. Flattery [39] also published tables of solutions in terms of the coefficients in series of associated Legendre functions for given frequencies of oscillation.

Kasahara [57] modified the algorithms of Longuet-Higgins and Flattery in such a way that the problem can be treated using a standard eigenvalue-eigenfunction routine. Swarztrauber and Kasahara [105] formulated the method consisting of expanding the eigenfunctions of L defined by (3.17) in terms of the spherical vector harmonics [86]. Computing softwares are available at the National Center for Atmospheric Research to calculate the eigenfrequencies and eigenfunctions of (3.20) for a positive value of ε.

An elementary solution $\mathbf{W}' = (u',v',z')^T$ of (3.6) is expressed by

$$\mathbf{W}'(\lambda,\phi,p,t) = S_n \, \mathbf{H}_\ell^s(\lambda,\phi;n) \, G_n(p) \exp(-i \, \sigma_\ell^s \, t), \qquad (3.22)$$

where $H_\ell^s(\lambda,\phi;n)$ denotes the horizontal structure of a normal mode, depending on zonal wavenumber s, meridional index ℓ, and vertical mode n, and it will be referred to as <u>Hough harmonics</u>. Also, σ_ℓ^s denotes the frequency of the normal mode as a function of three indices, s, ℓ, and n. Symbol S_n denotes the scaling matrix

$$\mathbf{S}_n = \begin{pmatrix} \sqrt{gD_n} & 0 & 0 \\ 0 & \sqrt{gD_n} & 0 \\ 0 & 0 & D_n \end{pmatrix}, \tag{3.23}$$

where D_n is the value of equivalent height corresponding to the nth vertical mode.

Hough harmonics are expressed by

$$H_\ell^s(\lambda,\phi;n) = \textcircled{H}_\ell^s(\phi;n)e^{is\lambda} \tag{3.24}$$

in which

$$\textcircled{H}_\ell^s(\phi;n) = \begin{pmatrix} U_\ell^s(\phi;n) \\ -i\ V_\ell^s(\phi;n) \\ Z_\ell^s(\phi;n) \end{pmatrix} \tag{3.25}$$

is the meridional structure function, called Hough vector function. The symbol i $[\equiv(-1)^{1/2}]$ in front of V_ℓ^s is introduced to deal with the phase shift of v' with respect to u' or z'. The minus sign is introduced so that the meridional distributions of U_ℓ^s, V_ℓ^s, and Z_ℓ^s agree with those given in [70].

Hough harmonics $\mathbf{H}_\ell^s$ are orthogonal in the sense that

$$(2\pi)^{-1} \int_0^{2\pi} \int_{-1}^{1} \mathbf{H}_{\ell'}^{s'} \cdot (\mathbf{H}_\ell^s)^* \, d\mu \, d\lambda = \delta_{\ell\ell'} \delta_{ss'}, \tag{3.26}$$

where $\mu = \sin\phi$, the asterisk in (3.26) denotes the complex conjugate, and the dot denotes the inner product. This orthogonality is shown from the property of the linearized shallow-water equations (3.16) [35, 93] for $s \geq 1$. Matrix $\mathbf{L}$ is skew Hermitian. Therefore, the frequency σ is real and the eigenfunctions corresponding to discrete eigenfrequencies are orthogonal [81, 73].

For $s \geq 1$, two kinds of normal modes exist: high-frequency eastward and westward propagating inertia-gravity waves and low-frequency westward propagating rotational waves of Rossby-Haurwitz type [49, 50, 97]. Both Margules [80] and Hough [53] found these two kinds of waves for small values of

ε and called high-frequency waves the first kind and low-frequency waves the second kind.

The case of s = 0 is unique in that the frequencies of gravity motions (first kind) appear as pairs of positive and negative values of the same magnitude and the frequencies of rotational motions (second kind) are all zero [70]. The Hough vector function of the first kind, $Ⓗ_\ell^0$ for negative frequency is the complex conjugate of $Ⓗ_\ell^0$ for positive frequency. Because the frequencies of gravity waves are all discrete, the normal modes of the first kind are orthogonal in the sense of (3.26). However, the rotational modes with zero frequencies are not necessarily orthogonal. In fact, it was not clear what was meant by the normal modes with zero frequency for s = 0. Kasahara [59] proposed a method for constructing a set of orthogonal vector functions for the second kind and called them geostrophic modes which, however, required orthogonalization in their formulation. Shigehisa [102] developed another method of solution to the zonal mode problem in which the zero frequency rotational modes of s = 0 are obtained as the limit of σ_r^s/s as $s \to 0$. This approach is attractive because these zonal modes exhibit the same fundamental characteristics of rotational modes for s > 0, including the property of orthogonality as defined by (3.26). Hence, with this set of rotational modes for s = 0, Hough harmonics are orthogonal and complete for $s \geq 0$. Shigehisa used a polynomial power expansion to calculate the rotational modes for s = 0. This computational approach, however, is not efficient. Swarztrauber and Kasahara [105] improved Shigehisa's approach by using vector spherical harmonic expansion and the calculation of these rotational modes is included in the software package mentioned earlier.

4. SOLUTIONS OF THE NONLINEAR SYSTEM BY NORMAL MODE EXPANSION

We express the dependent variables in the vector form

$$W(\lambda,\phi,p,t) = (u,v,z)^T. \tag{4.1}$$

Then, the general solution of (3.1)-(3.3) may be given by

$$\mathbf{W}(\lambda,\phi,p,t) = \sum_n S_n \mathbf{W}_n(\lambda,\phi,t)\, G_n(p), \qquad (4.2)$$

where S_n is the scaling matrix given by (3.23) and

$$\mathbf{W}_n(\lambda,\phi,t) = (u_n, v_n, z_n)^T \qquad (4.3)$$

denotes the dimensionless horizontal structure solution for the nth vertical mode.

By substituting (4.2) into (3.1)-(3.3), scaling time t by 2Ω, multiplying the resulting equations by $G_m(p)$, integrating them over the vertical domain from o to p_s, and utilizing the vertical structure equation (3.11) and the orthogonality condition (3.14), we obtain

$$\partial\mathbf{W}_n(\lambda,\phi,\hat{t})/\partial\hat{t} + \mathbf{L}_n\mathbf{W}_n(\lambda,\phi,\hat{t}) = \mathbf{N}_n(\lambda,\phi,\hat{t}), \qquad (4.4)$$

where $\mathbf{L}_n$ stands for operator $\mathbf{L}$ defined by (3.17), except for the equivalent height D being replaced by D_n corresponding to the nth vertical mode and

$$\mathbf{N}_n(\lambda,\phi,\hat{t}) = \begin{pmatrix} (2\Omega)^{-1}\,(gD_n)^{-1/2}\,P_n[N_u] \\ (2\Omega)^{-1}\,(gD_n)^{-1/2}\,P_n[N_v] \\ -\,(2\Omega)^{-1}\,P_n[N_z] \end{pmatrix} \qquad (4.5)$$

with P_n being the vertical projection operator defined by

$$P_n[\ \ ] = p_s^{-1} \int_o^{p_s} [\ \ ]\, G_n(p)\, dp. \qquad (4.6)$$

By the use of operator (4.6), the vertical dependence of (3.1)-(3.3) is separated out, leading to the nonlinear horizontal structure equation (4.4).

We again resort to a normal mode expansion and assume that

$$\mathbf{W}_n(\lambda,\phi,\hat{t}) = \sum_{r=1}^{R}\sum_{s=-M}^{M} W_r^s(\hat{t};n)\, \mathbf{H}_r^s(\lambda,\phi;n), \qquad (4.7)$$

where $\mathbf{H}_r^s(\lambda,\phi;n)$ denotes Hough harmonics defined by (3.24) and $W_r^s(\hat{t};n)$ is the expansion coefficient. The summation for serial number r from 1 to R should include all the meridional modes of the first and second kinds mentioned in Section 3.

Formally, R should be infinite as is the limit of zonal wavenumber M. In reality, the series must be truncated with finite limits of R and M. Since $\mathbf{W}_n$ is real, while W_r^s and $\mathbf{H}_r^s$ are complex, we impose the reality conditions

$$W_r^{-s} = (W_r^s)^* \quad \text{and} \quad \mathbf{H}_r^{-s} = (\mathbf{H}_r^s)^*, \tag{4.8}$$

where the asterisk denotes complex conjugate.

Substituting (4.7) into (4.4), integrating the resulting equation over the entire globe after multiplication by the complex conjugate of $\mathbf{H}_r^s$, utilizing the fact that H_r^s and $\hat{\sigma}_r^s$ satisfy the relationship

$$\mathbf{L}_n \mathbf{H}_r^s = i\, \hat{\sigma}_r^s\, \mathbf{H}_r^s, \tag{4.9}$$

and utilizing the definition (3.24) and the orthogonality condition (3.26), we obtain the spectral equation

$$dW_r^s(\hat{t};n)/d\hat{t} + i\hat{\sigma}_r^s\, W_r^s(\hat{t};n) = N_r^s(\hat{t};n), \tag{4.10}$$

where

$$N_r^s(\hat{t};n) = \int_{-1}^{1} \mathbf{N}_s(\phi,\hat{t};n)\cdot \mathbf{H}_r^{s*}\, d\mu \tag{4.11}$$

with $\mu = \sin\phi$, and

$$\mathbf{N}_s(\phi,\hat{t};n) = (2\pi)^{-1} \int_0^{2\pi} \mathbf{N}_n(\lambda,\phi,\hat{t})e^{-is\lambda}\, d\lambda. \tag{4.12}$$

Integral (4.12) is the Fourier transform of the nonlinear term and (4.11) may be called the Hough transform. We can evaluate (4.11) and (4.12) by using the transform method [37,87]. Within the accuracy of the truncated representation of $\mathbf{N}_n(\lambda,\phi,\hat{t})$, (4.12) and (4.11) can be evaluated exactly by using the Fourier transform and the Gaussian quadrature, respectively. Once N_r^s in (4.10) is evaluated, the ordinary differential equation (4.10) can be solved numerically with respect to time [58,59]. The calculation of the linear term with a constant coefficient $\hat{\sigma}_r^s$ may be performed without time truncation errors by redefining a new variable $R_r^s = W_r^s \exp(i\hat{\sigma}_r^s\hat{t})$ and extrapolating R_r^s in time [7].

5. NORMAL MODE INITIALIZATION

We have already mentioned that there are two kinds of normal modes, high-frequency gravity-inertia waves and low-frequency rotational waves. In the atmosphere, the low-frequency waves (second kind) are quasi-geostrophic and meteorologically significant, while the amplitudes of the high frequency waves (first kind) are, in general, very small. Hence, it is important to ensure that the amplitudes of the high-frequency gravity motions be small initially when the primitive equations are solved as an initial value problem. Otherwise, the large-amplitude gravity waves left in the initial condition, due to observational and analysis errors, will appear during the time integration and may overwhelm meteorologically significant motions. The process of adjusting the input data for the prediction model to ensure small amplitudes of gravity waves in the initial condition is called initialization. None of the initialization schemes developed up to the mid-1970's [13] can satisfactorily balance planetary-scale motions and apply to the tropics.

In recent years, a new approach called nonlinear normal mode initialization (NNMI) has been developed by Baer [6] and Machenhauer [74]. This approach, which is still under development, offers a promising solution to one of the long-standing problems in numerical weather prediction and has been adopted at many operational forecasting centers [2,29, 31,108,123].

In order to illuminate the essential features of NNMI, let us simplify the notation of spectral equations (4.10) as follows:

$$dW_G/dt + i\ \sigma_G\ W_G = N_G \qquad (5.1)$$

for the first kind and

$$dW_R/dt + i\ \sigma_R\ W_R = N_R \qquad (5.2)$$

for the second kind. The frequency of the first kind, indicated by σ_G, is generally one order of magnitude larger than σ_R for a not-so-small equivalent height D_n. The simplest initialization scheme is, therefore, to set $W_G = 0$ initially [36]. However, the elimination of all the spectral coefficients of the first kind in the initial condition still leads to generation of small-amplitude gravity waves due to the nonlinear term N_G in (5.1) [119]. Machenhauer's procedure [74] is to assume that the time derivative dW_G/dt be zero initially. This leads to the initial specification of W_G to satisfy

$$W_G = - i \, \sigma_G^{-1} \, N_G. \tag{5.3}$$

Since the nonlinear terms N_G contains W_G as well as W_R, an iteration scheme is used to solve the nonlinear equation (5.3), starting from an initial guess of W_G.

The convergence property of such an iteration scheme, however, is by no means clear and there are many studies related to this question [90, 10, 38, 109, 64]. A different approach to NNMI which does not require iteration to achieve a balanced state is described by Baer and Tribbia [8]. Their procedure is similar to an asymptotic expansion by the method of bounded derivative for the initialization of ordinary differential equations with different time scales proposed by Kreiss [65]. The Baer-Tribbia scheme is a successive correction method in which the zeroth-order estimate is $W_G = 0$. The first-order estimate of W_G is determined from (5.3) using the zeroth-order estimate on the right-hand side. The second-order estimate is obtained from the condition by setting d^2W_G/dt^2 to be zero initially which leads to

$$W_G = - i \, \sigma_G^{-1} \, N_G + \sigma_G^{-2} \, dN_G/dt. \tag{5.4}$$

The second-order estimate of W_G is then calculated by using the first-order estimate on the right-hand side. The higher the order of the time derivative which is bounded, the smoother the time evolution of the solution [65]. The higher-order estimate can be derived from the condition by setting the mth time derivative of W_G to be zero initially which leads to

$$W_G = - i \, \sigma_G^{-1} \, N_G + \sigma_G^{-2} \, dN_G/dt + \dots$$
$$- (i \sigma_G^{-1})^m \, d^{m-1} N_G/dt^{m-1}. \tag{5.5}$$

For the time integration of low order systems, the application of the higher-order initialization scheme is straightforward [6,72]. However, for a prediction system involving a large number of variables, the calculation of the time derivatives of N_G becomes time-consuming. Simplified methods of estimating the time derivatives of N_G have been proposed [111,114].

One practical question in applying the initial balance condition is to decide the extent of the normal mode coefficients which require initialization. Although the gravity waves (first kind) are generally considered to be meteorologically insignificant, some large-scale gravity waves such as the Kelvin and mixed-Rossby gravity waves and also gravity waves associated with small equivalent heights are important. Therefore, the application of initialization to those waves is harmful [95] and the initial conditions for those meteorologically significant waves must be specified as analyzed based on reliable observations. For the purpose of initialization, it is convenient to use a rather loose classification of "slow" modes for describing meteorologically significant motions and "fast" modes for the rest of the collection of normal modes. Leith [68] coined the word "slow manifold" for the set of normal modes with which the slow evolution of a model state can be described. For the large-scale motions of the atmosphere, the effects of heating/cooling and energy dissipation in the boundary layer are important. Hence, the set of normal modes which is needed to define the slow manifold in a realistic model atmosphere may differ considerably from that of an adiabatic and inviscid model atmosphere.

A reason why the use of normal mode expansion for initialization is attractive is its flexibility in selection of the slow manifold. On the other hand, the interpretation of the balance procedure in light of the model dynamics in physical space is difficult, unless the number of normal modes involved in the model atmosphere is small. A question has been raised concerning the relationship between the NNMI and the balance procedures based on quasi-geostrophic theory [27,

89] in connection with the non-convergence of Machenhauer's iteration in the non-elliptic regions of the traditional non-linear balance equation [27,17] when the geopotential height is held unchanged during the iteration procedure [28,112]. The equivalence between the classical balance procedure by quasi-geostrophic theory and the first-order estimate (5.3) has been demonstrated for mid-latitude baroclinic models whose normal modes are obtainable from a linearized constant Coriolis parameter model with a basic state at rest [68,62].

6. OBSERVATIONAL EVIDENCE OF PLANETARY-SCALE WAVES

Large-scale weather changes in the mid-latitudes are primarily controlled by mid-tropospheric long waves, with wavelengths of several thousand kilometers, in the westerlies. After the pioneering work of C. G. Rossby [97] and others, these large-scale disturbances are called Rossby waves. The observational and theoretical investigations on the characteristics of these waves have been central to modern meteorology since the invention of radiosondes for measuring meteorological elements in the free atmosphere. The propagation of Rossby waves is approximately governed by the conservation law of absolute vorticity (a vorticity relative to the rotating earth plus the vorticity of the earth's own rotation around the vertical) in an inviscid and nondivergent fluid.

It is now well established that large-scale Rossby waves are very similar in nature to the rotational waves (second kind) corresponding to the external mode [78]. Rossby treated the normal mode problem using a beta-plane in which the meridional variation of the Coriolis parameter is approximated by a constant which is now referred to as Rossby parameter. Haurwitz [50] considered an extension of Rossby's work in spherical geometry so that the meridional variation of the Coriolis parameter, as well as the curvature of flow, are correctly accounted for. Haurwitz waves are identical to the rotational waves corresponding to an infinite equivalent height ($D \to \infty$).

On the third column in Table 1, the angular velocities in degrees longitude per day of Haurwitz waves for zonal wavenumber s = 1 are shown. The minus sign indicates westward propagation. The meridional modal index is shown on the first column. The gravest meridional mode is denoted by ℓ = 0 for which the height field is antisymmetric (indicated by A in the second column) about the equator. The next gravest mode is denoted by ℓ = 1 for which the height field is symmetric (indicated by S) about the equator. The fourth column shows the phase velocities of rotational waves in degrees longitude per day for the equivalent height D = 10 km. It is seen that Haurwitz waves propagate westwards with much greater speeds than those of rotational waves for s = 1. The fifth to seventh columns show the phase velocities of rotational waves for s = 2 to 4. The phase speeds become small for increased zonal wavenumber s and meridional index ℓ.

Table 1. Propagation phase speeds in degrees longitude per day of Haurwitz waves, $D \to \infty$, and rotational waves of second kind, D = 10 km for zonal wavenumber s = 1-4. Numbers in parentheses indicate periods of free oscillations in days. Reference state is an atmosphere at rest.

ℓ	P_a	H.W. $D=\infty$ s=1	rotational waves D = 10 km s=1	s=2	s=3	s=4
0	A	-360	-303(1.2)	-111(1.6)	-57(2.1)	-35(2.6)
1	S	-120	-72(5.0)	-48(3.7)	-32(3.7)	-23(4.0)
2	A	-60	-43(8.3)	-30(5.9)	-21(5.5)	-16(5.6)
3	S	-36	-29(12.3)	-21(8.5)	-16(7.6)	-12(7.4)
4	A	-24	-21(17.3)	-16(11.5)	-12(10.0)	-10(9.4)

ℓ ≡ meridional modal index

P_a : A ≡ antisymmetric; S ≡ symmetric.

The presence of discrete free rotational waves has been detected in the troposphere for those modes listed above the dashed line [78,1] except those corresponding to $\ell = 0$. However, the mode ($s = 3$, $\ell = 0$) which has a period of 2.1 days is detected in the middle atmosphere [98]. Considerable attention has been paid recently to the modes ($s = 1$, $\ell = 1$) and ($s = 1$, $\ell = 3$) and there are many investigations on these modes. See the references of [78,61]. Because planetary waves propagate in the prevailing westerlies, the wave phase velocities are modified by the basic zonal flow and its associated dynamics. If we denote the maximum and minimum basic state zonal velocities by $\bar{u}_{max}$ and $\bar{u}_{min}$ and the phase velocity of wave by c, then the waves satisfying the conditions that $c > \bar{u}_{max}$ and $c < \bar{u}_{min}$ appear as discrete modes. The eastward and westward propagating gravity waves (first kind) with D = 10 km easily satisfy these conditions. Most of the rotational waves (second kind), however, except those above the dashed line in Table 1, may fall in the range $\bar{u}_{min} \leq c \leq \bar{u}_{max}$ and they will not appear as discrete modes. The periods of oscillations are also affected by the presence of the basic flow. For example, the mode ($s = 1$, $\ell = 3$) has a period of 12.3 days without basic flow, but the period of this mode becomes about 17 days with the effect of basic flow [61].

Since the planetary waves possess large kinetic energy and act as the steering flows for synoptic weather systems, it is important to specify accurately these waves in the initial conditions for forecasting. Scarcity of meteorological observations in the southern hemisphere and in the tropics tends to produce incorrect analysis for these global-scale motions. Because their phase velocities are relatively large, the incorrect initial specification of these scales of motion can contribute to substantial forecast errors in the middle to higher latitudes in a few days [103,33].

7. NUMERICAL METHODS FOR SOLVING GLOBAL ATMOSPHERIC MODELS

Numerical integration methods for solving atmospheric prediction equations have been discussed in many literatures [66, 84, 79, 60, 48, 24]. Here, I will address myself only to the current methods for solving global atmospheric models. In the past, three approaches were considered to apply finite-difference approximations to fluid flow over a sphere, namely the use of map projections, quasi-homogeneous grids, and longitude-latitude grids. The most commonly used grid today is a uniform latitude-longitude grid. Difference approximations are written over grid points defined by the intersections of latitude and longitude circles. However, the latitude-longitude grid system too has a number of computational problems. Poles are singular where wind components in spherical coordinates are undefined. The maximum allowable time step with an explicit time integration scheme is reduced proportionally due to the convergence of meridians near the poles. A number of expediencies have been tried to overcome this computational difficulty. The most practical solution appears to be application of a Fourier filtering in the longitudinal direction to relax the CFL stability condition of explicit schemes [120, 3, 115, 23, 55, 83].

For the primitive equation models, a short time step with explicit schemes is dictated by fast-moving gravity waves with a speed of about 300 m sec^{-1} rather than meteorologically significant waves with one order of magnitude less speed. One technique to economize the explicit integrations involving different time-scale phenomena is known as the method of fractional steps or splitting. The idea is to solve a multi-time scale prediction problem by dividing it into a series of steps, containing difference operations of only one dimension or some parts of the differential equations involved [79]. By this technique, the horizontal advection terms in the prediction equations are integrated with a time step limited by the advection speed, while the terms which describe gravity waves are integrated in a succession of shorter steps [43]. Since the gravity waves associated with small equivalent heights move with almost the same speeds of meteorologically significant waves, the partition of "fast" and "slow" modes can be performed by calculating the phase speeds in different vertical modes. Then the "fast" modes are integrated explicitly and separately, with time steps allowed by the respective CFL conditions and are recombined at periodic time intervals. The remaining "slow" modes are integrated explicitly with a time step larger than that allowed for the fastest moving waves [77, 25].

When the time step of an explicit scheme is required to be shorter than that desired by the consideration of accuracy, an obvious solution to avoid using a short time step is the application of implicit method. A full treatment of the implicit method, however, is time consuming in general, though an efficient algorithm may be developed [12]. An alternative is to apply an implicit procedure only to the "fast" waves, whereas the "slow" modes are calculated explicitly. This approach is called a semi-implicit scheme [96]. The explicit version of the ECMWF finite-difference operational forecast model uses a time step of 5 minutes for a 1.875° latitude-longitude grid [23]. A semi-implicit version of the same model uses a time step of 15 minutes [51].

While the finite-difference method has been used extensively in the past, the spectral approach [75] is gaining more popularity in modeling of the global atmosphere. The spectral form of the barotropic nondivergent vorticity equation over a sphere has been formulated with spherical harmonics as the basis functions [92]. The advantages of this procedure are 1) a mapping of the sphere is not necessary, so that the difficulties with finite-difference methods near the poles do not arise, 2) the truncated spectral equations possess the conservation properties of energy and enstrophy and these will stabilize the time-dependent solutions of spectral equations, and 3) spherical harmonics are eigensolutions of a linearized version of the vorticity equation so that the linear part of the equation can be treated exactly.

During the 1960s, the spectral equations of atmospheric models for more general flows than two-dimensional nondivergent motions were formulated using spherical harmonics as the basis functions. These spectral equations involve a large number of nonlinear interaction terms and, therefore, require a substantial amount of arithmetic. In 1970, Eliasen *et al.* [37] and Orszag [87] developed the transform method to speed up the evaluation of the spectral contribution of nonlinear interaction terms which involves the integration of nonlinear terms with respect to the spatial independent variables. The essence of the transform method is already discussed in Section 4. An additional feature of the transform method over the traditional interaction coefficient method is the ease of treating the physical processes in the prediction model since they are calculated in the physical space.

For formulating the spectral model of the prediction equations presented in Section 2, it is customary to transform Eqs. (2.2) and (2.3) into the equations dealing with the time rate of change of relative vorticity $\zeta = \mathbf{K} \cdot \nabla \times \mathbf{V}$, where $\mathbf{K}$ denotes the vertical unit vector, and horizontal divergence $\delta = \nabla \cdot \mathbf{V}$. Then, the stream function ψ and the velocity potential χ are introduced to represent the horizontal velocity by $\mathbf{V} = \mathbf{K} \times \nabla\psi + \nabla\chi$, vorticity by $\nabla^2\psi$ and divergence by $\nabla^2\chi$. Dependent variables ψ, χ, temperature T, geopotential height z, vertical velocity ω are represented in terms of series in spherical harmonics $P_n^s(\sin\phi)\exp(is\lambda)$. The expansion coefficients are functions of vertical coordinate p and time t. Time rates of change in the spectral coefficients are determined on the basis of the Galerkin approximation [75]. The discretization of the spectral coefficients in the vertical is done almost exclusively by the finite-difference method. The placement of dependent variables and vertical differencing schemes is often selected by requiring that the discretized equations satisfy analogues of some important properties of the differential equations [4]. One advantage in the spectral representation in the horizontal is that the implementation of the semi-implicit algorithm is much simpler and more efficient in the spectral models than in the finite-difference models.

Current global spectral models for research and operational forecasting [52, 34, 19, 5, 101, 46, 121] are patterned after Bourke [18] with various modifications and improvements. Detailed reports [54,45] were published from the European Centre for Medium Range Weather Forecasts (ECMWF) concerning the performance comparison between the ECMWF finite-difference operational model [51] and two different horizontal resolution versions of the global spectral model [5]. Table 2 shows the timing comparison of two different models [54]. A comparative study [54] based on 53 cases of 10-day forecasts with the T63 spectral and 1.875° grid-point model shows that the T63 spectral model gives significantly better results. Apparently, superiority of the T63 spectral model comes from the reduction of propagation phase errors of synoptic systems by improvement in the calculation of non-linear advection terms. Since the computing time required to run both models is comparable, the T63 spectral model may be advantageous over the 1.875° grid-point model.

Table 2. CRAY1 time in minutes to run the models [54]

	Grid-point model	Spectral model	
		T63	T40
CPU time	160	165	45
CPU spent in the system	20	20	5
Waiting for I/O	20	25	15
Post-processing	30	30	25
TOTAL	230	240	90

Model description: Both models have 15 vertical levels.

Grid-point model (1.875° grid resolution)

97 latitude lines, including the poles, and 192 points on each line. Δt = 15 min.

Spectral model

T40: Triangular wavenumber 40 truncation

64 Gaussian latitude lines with 128 points on each line. Δt = 26 min. 40 sec.

T63: Triangular wavenumber 63 truncation

96 Gaussian latitude lines with 192 points on each line. Δt = 18 min.

Now, it may be worthwhile to consider possible areas of further improvement in the numerical formulation of dynamical models. One area is the discretization in the vertical. At present, advanced forecasting models have a typical vertical resolution of about 100 mb in the troposphere with several additional levels near the surface and in the stratosphere. Although centered differencing is always used, the accuracy of vertical differencing is only of the first order due to the nonuniformity of the vertical grids. Despite rather successful application of the spectral approach to the representation of variables in the horizontal, only limited attempts have been made to represent the vertical model structure by using empirical orthogonal functions [16], Legendre polynomials [76], and a finite-element formulation [104]. I have discussed in Section 4 the representation of variables in terms of the model normal modes in which the vertical expansion functions are derived from solutions to the vertical structure equation. These vertical modes may be "optimal" in the sense that they can be determined by a Galerkin procedure. Since the vertical structure of the large-scale motions is less known than the horizontal structure, it is necessary to examine the characteristics of atmospheric vertical structure before application of the normal mode expansion in the vertical.

In Section 4, I have discussed the use of Hough harmonics as the basis functions to formulate a spectral model of the global baroclinic primitive equations. Advantages of using Hough harmonics for a spectral global primitive equation model are that the prognostic variables are efficiently represented and the linear part of the prediction model can be treated exactly, since the Hough harmonics are normal modes of the prediction model. The initialization procedure is no longer separated from the time integration scheme, but it is a part of the time evolution calculation. In fact, one can formulate various time integration schemes depending on the selection of "slow" modes for explicit time integrations and of "fast" modes for diagnostic balancing in the same way as in the initialization procedure discussed in Section 5. It is no longer necessary to invoke the semi-implicit procedure, since the continuous application of the initialization procedure for the fast modes enables us to take a longer time step which is dictated only by the time integration of the slow modes [30]. Test calculations with the global shallow-water equations indicate that the Hough spectral approach is an efficient, accurate, and viable numerical method for the global model [58, 59, 9]. On the other hand, there is a computational disadvantage at present due to the need of storing internally a large number of Hough function tables for various wave components and vertical modes of the baroclinic model. Future developments in computer technology will easily alleviate this computational deficiency.

A question to ponder in atmospheric modeling today is why all numerical prediction models behave similarly to each other rather than to the real atmosphere. All numerical models exhibit systematic forecast errors which are similar to each other when a sufficient number of cases are considered. For example, it is reported in [15] that the forecast error signatures of the operational finite-difference ECMWF model and the National Meteorological Center's operational spectral model during the two winters of 1980-81 and 1981-82 were found to be very similar. Since the numerics of the two models are very different, the similarity of systematic errors in the two models is caused by other factors than the difference in the use of spectral and finite-difference methods for the horizontal discretization. Wallace *et al*. [118] reported that the introduction of more realistic orography in the ECMWF grid-point model results in a significant reduction of the systematic errors beyond 4 days, though the enhancement of the orography slightly degrades the shorter range forecasts. It is reported in [45] that the time evolutions of the globally-averaged kinetic energy and available potential energy in the ECMWF model are deficient compared with those of the real atmosphere, partly due to the deficiency in the global average heating rate. It is possible that the introduction of "higher" mountains may have increased the meridional heat transport to offset the net cooling trend and, hence, had a beneficial effect on medium-range forecasts.

It is important for a forecast model to maintain its own statistical equilibrium state which ideally is the same as the real climate. The tendency of a forecast model to seek its own statistical equilibrium state different from the real atmosphere is referred to as *model climate drift*. Clearly all the physical processes in the prediction models must be accurately parameterized. This indicates that medium-range forecast models should also be realistic climate models. The present computing capabilities are inadequate for performing various climate sensitivity studies with the state-of-the-art forecasting models. It may not be too far in the future, however, when the seemingly independent numerical efforts for weather prediction and climate simulation will eventually merge into a single endeavor.

8. OBJECTIVE ANALYSIS OF METEOROLOGICAL DATA

Besides advances in atmospheric modeling and computer capabilities, credit for recent forecast improvements goes to additional observations using the unconventional platforms such as aircrafts and constant-level balloons, cloud drift winds from geostationary satellites, and infrared and microwave temperature soundings which have been implemented in preparation for the First GARP Global Experiment [14,47]. The handling of this vast amount of atmospheric data has also revolutionized the methods of meteorological data analysis.

The automatic procedure for transforming observed data from geographic (and irregularly spaced) locations at various times to numerical values at the prescribed grid points at fixed time is called objective analysis. The most popular approach today, called optimum interpolation, involves the linear combination of the departure of observed data from the corresponding forecast values in such a way as to minimize the mean-square interpolation error within a certain region centered around a particular grid point of analysis [82,100, 71]. One advantage of this technique compared with the polynomial fitting is that the accuracy of observational data and their initial guesses (predicted values) can be taken into account by multivariate statistical analysis. Clearly the quality of the analysis depends on the accuracy of forecasts as well as the reliability of observations.

Besides the presence of observational errors, the basic problem of objective analysis arises from the fact that not all the dependent variables of a forecasting model are observed at a given location and time. Since measurements are carried continuously in time, can we infer the complete state of the atmosphere by using the time-sequence data of a limited number of meteorological observations? Suppose an observing system is capable of measuring only temperatures, but not winds. Is it possible to infer the wind field from the historical temperature data? The process of periodically updating observed data in place of predicted values during the time integration of a prediction model in order to depict the complete atmospheric state is called four-dimensional data assimilation. See references in [13].

Theoretical questions have been investigated under which the process of four-dimensional assimilation with incomplete but accurate data would define uniquely the complete atmospheric state [106,22]. As long as inserted data are accurate, a large number of numerical experiments with general circulation models have indicated that the reconstruction of the atmospheric model state appears feasible by the four-dimensional data assimilation process [13]. The effectiveness of the data assimilation is judged by two measures. One is the asymptotic error level which is the root-mean-square difference between the inserted data and the corresponding model-generated values after a long period of data insertion. The other is the rate of convergence to the asymptotic error level. Both measures depend on the discrepancy between the inserted data and the model-generated values, the type and geographical distribution of inserted data, and the frequency and technique of data insertion.

The discrepancy between the inserted data and the model-generated values is referred to as model shock and many techniques were developed in the past to reduce the amount of shock generated in the prediction models [85]. An important result from a large number of numerical experiments is that the asymptotic error level would become smaller and the rate of convergence faster if both the wind and mass data, which are mutually balanced under the model dynamics, are updated simultaneously even when only one of the observations is available for updating. See Section 4.2.1 in [13].

Without getting into a detailed discussion concerning the mechanism of assimilation dynamics, the above finding simply indicates the need of initialization for the inserted data. An "ideal" data insertion technique is looked upon as a procedure whereby the inserted data can stay on the model's slow manifold without generating "fast" modes [32].

The realization that the concept of initialization discussed in Section 5 is also relevant to understanding objective analysis pinpoints a source of deficiency which existed in past practice of numerical weather prediction; namely objective analysis and initialization have been treated independently as two separate steps. Before the era of numerical prediction, meteorological analysis was guided by the faithful visualization of observed data in terms of weather maps, assisted only by synoptic models and analysts' experiences. This analysis principle lasted until recently despite the fact that analysts had begun to use numerical forecasts for guidance in their analysis process. The necessity to change this old practice came from the realization that sometimes analysis products had to be changed at the initialization step. In some cases, the initialization procedure forced changes in the analysis that were larger than observational errors, even over data-rich regions!

The use of variational techniques coupled with the initialization procedure reduces some of the above discrepancies. Variational techniques are designed to minimize differences between the original and adjusted analyses in a least-square sense with or without additional dynamical constraints [99]. In these techniques, the accuracies of wind and mass observations can be taken into account through specified weight functions. The use of variational techniques in conjunction with the nonlinear normal mode initialization, applicable to the global barotropic model, was developed by Daley [28]. Because the dynamical constraint is nonlinear, he used a two-step process. First, a variational scheme was set up to determine "optimum" rotational components of the mass and wind fields with appropriate variational weights. Then, the gravity mode amplitudes were calculated using the nonlinear normal mode balancing scheme. A new variational integral was then set up by using the gravity mode projections as a strong constraint through Lagrange multipliers. This two-step process was repeated to obtain progressively better balanced fields. Daley's variational method was recently improved by Tribbia [113] by formulating the variational problem without using Lagrange multipliers for the initialization constraints and by substantially reducing the amount of computation. The explicit use of the Hough harmonic expansion for the mass and wind fields simplifies the algebra involved in the minimization of the variational integral. The spectral equations governing the expansion coefficients for the adjusted mass and wind fields are nonlinear, because the initialization constraints are nonlinear. Their solutions are obtained by linearization and by an asymptotic expansion or successive approximation. Although Tribbia demonstrated application of his method to the global two-dimensional analysis, the same formalism may be adapted to the global three-dimensional analysis using the three-dimensional normal mode expansion discussed in Section 4 with additional computation.

An alternative to the variational analysis-initialization approach is proposed by Williamson and Daley [122]. It consists of iterating multivariate optimum interpolation and nonlinear normal mode initialization. While the previous variational approach aims at minimizing the difference between a given analysis and the adjusted balanced state, this approach is designed to minimize the difference between the balanced state and the observations themselves rather than an analysis of these observations. Phillips [91] proposes yet another approach to the same objective. He suggests analyzing, with the use of optimum interpolation, only the slow mode fields and calculating the covariance structure functions based on the structure of slow normal modes. The fast mode components of mass and wind are computed diagnostically through nonlinear normal mode initialization. Although the approach of Phillips may not be practical to implement in the operational environment, his argument illuminates the complex nature of the analysis-initialization combination.

The choice of either optimum interpolation or the variational approach for objective analysis may be a matter of preference rather than involved principle, since both the approaches are based on the same principle of minimization between the observed data and the analysis. Both approaches, however, have been developed more or less independently in the past. The variational method is being further generalized by incorporating various improvements and constraints [116, 117]. In an effort to improve the optimum interpolation approach, interest has recently been revived in the adaptation of the sequential estimation theory of stochastic dynamics systems and the use of the Kalman filter [44]. An effort to close the gap between the variational method and the optimum interpolation approach is also seen in [116].

One distinction between the optimum interpolation approach and the variational method applied in meteorology is that the former is a succession of local procedure, while the latter is a global one. If the initialization scheme is global as in the case of nonlinear normal mode balancing, the iteration of optimum interpolation and nonlinear normal mode initialization is a mix of local and global procedures. If one develops an initialization scheme which is local [21], such an initialization scheme would fit more naturally to the optimum interpolation procedure. Such a unified objective analysis-initialization scheme would be best suited to numerical prediction with a limited-area fine-mesh forecasting model.

Many operational forecasting centers have now adopted the optimum interpolation approach for global analysis. However, as long as the nonlinear normal mode balancing discussed in Section 5 is used as the initialization procedure, the variational approach with Hough harmonics as the basis functions and with judicious dynamical constraints would be more naturally suited to global analysis and forecasting. A rudimental form of such an analysis scheme was pioneered by Flattery [40]. The key to whether this type of analysis-initialization procedure is successful depends on how closely one can describe the general nonlinear balances of a realistic model atmosphere, with diabatic processes included, in terms of the three-dimensional normal modes.

9. CONCLUSIONS

Recent improvements in short- to medium-range weather prediction have appeared as the culmination of many efforts to improve meteorological observations, objective analysis, numerical approximations and physical parameterizations in parallel with the advance in electronic computers' capacity and speed. These efforts were direct results of the planning for participation in the Global Weather Experiment (GWE) in 1979 as the largest scientific experiment ever attempted to observe the atmosphere, utilizing every means of the observing systems. The GWE is also termed the First GARP Global Experiment (FGGE). GARP is the acronym for the Global Atmospheric Research Program, and is supported jointly by the World Meteorological Organizaiton (WMO) and the International Council of Scientific Unions (ICSU). The GWE data will be the best meteorological data set yet available for the next five to ten years.

In this essay, I have reviewed recent modeling and computing developments in the field of numerical weather prediction. The process of numerical prediction consists of four aspects - observations, analysis, initialization, and numerical modeling. In the past, these four aspects have been developed more or less independently by specialists in each area and the gaps have existed in the interfaces. Recognition of these existing gaps and the many efforts to eliminate a mismatch of the interfaces are clearly evident in recent developments. I have made an attempt to connect the three aspects - modeling, initialization, and objective analysis - using the concept of three-dimensional normal mode functions as a thread through the interfaces. The realization of my proposed dynamical system of numerical prediction, however, demands a high-speed computer which is faster and has a larger core memory than the present supercomputers of the CRAY-1 and Cyber 205 class.

ACKNOWLEDGMENTS

This research was conducted at the National Center for Atmospheric Research (NCAR) which is supported by the National Science Foundation. The original manuscript was reviewed by J. Tribbia and D. Williamson and their useful comments were incorporated in this manuscript. Partial support has been provided through the National Oceanic and Atmospheric Administration under P.O. No. NA83AAG00837.

REFERENCES

1. Ahlquist, J. E., 1982: Normal-mode global Rossby waves: Theory and observations. J. Atmos. Sci, 39, 193-202.
2. Andersen, J. H., 1977: A routine for normal mode initialization with non-linear correction for a multi-level spectral model with triangular truncation. Inter. Rep. No. 15, European Centre for Medium Range Weather Forecasts, Reading, England.
3. Arakawa, A. and V. R. Lamb, 1977: Computational design of the basic dynamical processes of the UCLA general circulation model. Methods in Comp. Phys., 17, 173-265.
4. Arakawa, A. and M. J. Suarez, 1983: Vertical differencing of the primitive equations in sigma coordinates. Mon. Weather Rev., 111, 34-45.
5. Baede, A.P.M., M. Jarraud and U. Cubasch, 1979: Adiabatic formulation and organisation of ECMWF's spectral model. Tech. Rep. No. 15, European Centre for Medium Range Weather Forecasts, Reading, England, 39 pp.
6. Baer, F., 1977: Adjustment of initial conditions required to suppress gravity oscillations in nonlinear flows. Beitr. Phys. Atmos., 50, 350-366.
7. Baer, F. and G. W. Platzman, 1961: A procedure for numerical integration of the spectral vorticity equation. J. Meteor., 18, 393-401.
8. Baer, F. and J. J. Tribbia, 1977: On complete filtering of gravity modes through nonlinear initialization. Mon. Weather Rev., 105, 1536-1539.
9. Baer, F. and J. J. Tribbia, 1983: On filtering of gravity modes in atmospheric models. Submitted to Mon. Weather Rev.

10. Ballish, B. A., 1981: A simple test of the initialization of gravity modes. Mon. Weather Rev.,109, 1318-1321.

11. Baumhefner, D., 1983: The relationship between present large-scale forecast skill and new estimates of predictability error growth. Proc. the Workshop on the Predictability of Fluid Motions, La Jolla Institute, La Jolla, CA.

12. Beam, R. W. and R. F. Warming, 1976: An implicit finite difference algorithm for hyperbolic systems in conservation form. J. Comp. Phys., 22, 87-110.

13. Bengtsson, L., 1975: Four-dimensional assimilation of meteorological observations. GARP Publ. Series No. 15, Global Atmospheric Research Programme, WMO-ICSU,Geneva, Switzerland, 76 pp.

14. Bengtsson, L., M. Kanamitsu, P. Kallberg, and S. Uppala, 1982: FGGE research activities at ECMWF. Bull. Amer. Meteor. Soc., 63, 277-303.

15. Bettge, T. W., 1983: A systematic error comparison between the ECMWF and NMC prediction models. Submitted to Mon. Weather Rev.

16. Bodin, S., 1974: The use of empirical orthogonal functions in quasi-geostrophic numerical prediction models. Tellus, 26, 582-593.

17. Bolin, B., 1955: Numerical forecasting with the barotropic model. Tellus, 7, 27-49.

18. Bourke, W., 1974: A multi-level spectral model. I. Formulation and hemispheric integrations. Mon. Weather Rev., 102, 687-701.

19. Bourke, W., B. McAvaney, K. Puri and R. Thurling, 1977: Global modeling of atmospheric flow by spectral methods. Methods in Comp. Phys., 17, 267-324.

20. Bowditch, N., 1832: Translation of "Mechanique Céleste" by P. S. Laplace, Vol. II, Chelsea Publ., New York (1966).

21. Browning, G., A. Kasahara and H.-O. Kreiss, 1980: Initialization of the primitive equations by the bounded derivative method. J. Atmos. Sci., 37, 1424-1436.

22. Bube, K. P. and M. Ghil, 1981: Assimilation of asynoptic data and the initialization problem. Dynamic Meteorology: Data assimilation methods, Springer-Verlag, New York, 111-138.

23. Burridge, D. M. and J. Haseler, 1977: A model for medium range weather forecasting. Tech. Rep. No. 4, European Centre for Medium Range Weather Forecasts, Reading, England, 45 pp.

24. Chang, J. (Ed), 1977: General circulation models of the atmosphere. Methods in Comp. Phys., 17, 332 pp.

25. Chao, W. C., 1982: Formulation of an explicit-multiple-time-step time integration method for use in a global primitive equation grid model. Mon. Weather Rev., 110, 1603-1617.

26. Chapman, S. and R. S. Lindzen, 1970: Atmospheric tides - thermal and gravitational, Gordon & Breach, New York, 200 pp.

27. Charney, J., 1955: The use of the primitive equations of motion in numerical prediction. Tellus, 7, 22-26.

28. Daley, R., 1978: Variational non-linear normal mode initialization. Tellus, 30, 201-218.

29. Daley, R., 1979: The application of non-linear normal mode initialization to an operational forecast model. Atmosphere-Ocean, 17, 97-124.

30. Daley, R., 1980: The development of efficient time integration schemes using model normal modes. Mon. Weather Rev., 108, 100-110.

31. Daley, R., 1981: Normal mode initialization. Rev. Geophys. Space Phys., 19, 450-468.

32. Daley, R. and K. Puri, 1980: Four-dimensional data assimilation and the slow manifold. Mon. Weather Rev., 108, 85-99.

33. Daley, R., J. Tribbia and D. L. Williamson, 1981: The excitation of large-scale free Rossby waves in numerical weather prediction. Mon. Weather Rev., 109, 1836-1861.

34. Daley, R., C. Girard, J. Henderson and I. Simmonds, 1976: Short-term forecasting with a multi-level spectral primitive equation model, Part I: Model formulation. Atmosphere, 14, 98-134.

35. Dickinson, R. E., 1966: Propagators of atmospheric motions. Ph.D. thesis, Dept. Meteorology, Massachusetts Institute of Technology, 243 pp.

36. Dickinson, R. E. and D. L. Williamson, 1972: Free oscillations of a discrete stratified fluid with application to numerical weather prediction. J. Atmos. Sci., 29, 623-640.

37. Eliasen, E., B. Machenhauer, and E. Rasmussen, 1970: On a numerical method for integration of the hydrodynamical equations with a spectral representation of the horizontal fields. Inst. Theoretical Meteorology, Copenhagen Univ., Denmark, Rep. No. 2, 35 pp.

38. Errico, R. M., 1983: Convergence properties of Machenhauer's initialization scheme. Submitted to Mon. Weather Rev.

39. Flattery, T. W., 1967: Hough functions. Tech. Rep. 21, Dept. Geophys. Sci., University of Chicago, 175 pp.

40. Flattery, T. W., 1970: Spectral models for global analysis and forecasting. Proc. the Sixth AWS Technical Exchange Conference, Air Weather Service Tech. Rep. 242, 42-53.

41. Fleming, R. J., T. M. Kaneshige and W. E. McGovern, 1979: The Global Weather Experiment I. The observational phase through the First Special Observing Period. Bull. Amer. Meteor. Soc., 60, 649-659.

42. Fleming, R. J., T. M. Kaneshige, W. E. McGovern and T. E. Bryan, 1979: The Global Weather Experiment II. The Second Special Observing Period. Bull. Amer. Meteor. Soc., 60, 1316-1322.

43. Gadd, A. J., 1978: A split explicit integration scheme for numerical weather prediction. Quart. J. R. Meteor. Soc., 104, 569-582.

44. Ghil, M., S. Cohn, J. Tavantzis, K. Bube and E. Isaacson, 1981: Application of estimation theory to numerical weather prediction. Dynamic Meteorology: Data Assimilation Methods, Springer-Verlag, New York, 139-224.

45. Girard, C. and M. Jarraud, 1982: Short and medium range forecast differences between a spectral and grid point model. An extensive quasi-operational comparison. Tech. Rep. No. 32, European Centre for Medium Range Weather Forecasts, Reading, England, 176 pp.

46. Gordon, C. T. and W. F. Stern, 1982: A description of the GFDL global spectral model. Mon. Weather Rev., 110, 625-644.

47. Halem, M., E. Kalnay, W. E. Baker and R. Atlas, 1982: An assessment of the FGGE satellite observing system during SOP-1. Bull. Amer. Meteor. Soc., 63, 407-426.

48. Haltiner, G. J. and R. T. Williams, 1980: Numerical prediction and dynamic meteorology: John Wiley, New York, 477 pp.

49. Haurwitz, B., 1937: The osillations of the atmosphere. Gerlands Beiträge zur Geophysik, 51, 195-233.

50. Haurwitz, B., 1940: The motion of atmospheric disturbances on the spherical earth. J. Marine Res., 3, 254-267.

51. Hollingsworth, A., K. Arpe, M. Tiedtke, M. Capaldo and H. Savijärvi, 1980: The performance of a medium-range forecast model in winter -- Impact of physical parameterizations. Mon. Weather Rev., 108, 1736-1773.

52. Hoskins, B. J. and A. J. Simmons, 1975: A multi-layer spectral model and the semi-implicit method. Quart. J. Roy. Meteor. Soc., 101, 637-655.

53. Hough, S. S., 1898: On the application of harmonic analysis to the dynamical theory of the tides - Part II. On the general integration of Laplace's dynamical equations. Phil. Trans. Roy. Soc. London, A 191, 139-185.

54. Jarraud, M., C. Girard and U. Cubasch, 1981: Compairson of medium range forecasts made with models using spectral or difference techniques in the horizontal. Tech. Rep. No. 23, European Centre for Medium Range Weather Forecasts, Reading, England, 95 pp.

55. Kalnay-Rivas, E. and D. Hoitsma, 1979: The effect of accuracy, conservation and filtering on numerical weather forecasting. Proc. of Fourth Conference on Numerical Weather Prediction, Amer. Meteor. Soc., 302-312.

56. Kasahara, A., 1974: Various vertical coordinate systems used for numerical weather prediction. Mon. Weather Rev., 102, 509-522.
57. Kasahara, A., 1976: Normal modes of ultralong waves in the atmosphere. Mon. Weather Rev., 104, 669-690.
58. Kasahara, A., 1977: Numerical integration of the global barotropic primitive equations with Hough harmonic expansions. J. Atmos. Sci., 34, 687-701.
59. Kasahara, A., 1978: Further studies on a spectral model of the global barotropic primitive equations with Hough harmonic expansions. J. Atmos. Sci., 35, 2043-2051.
60. Kasahara, A. (Ed), 1979: Numerical methods used in atmospheric models: Vol. 2. GARP Publ. Series No. 17, WMO-ICSU, Geneva, Switzerland, 499 pp.
61. Kasahara, A., 1980: Effect of zonal flows on the free oscillations of a barotropic atmosphere. J. Atmos. Sci., 37, 917-929. Corrigendum, J. Atmos. Sci., 38, 2284-2285.
62. Kasahara, A., 1982: Nonlinear normal mode initialization and the bounded derivative method. Rev. Geophys. Space Phys., 20, 385-397.
63. Kasahara, A. and K. Puri, 1981: Spectral representation of three-dimensional global data by expansion in normal mode functions. Mon. Weather Rev., 109, 37-51.
64. Kitade, T., 1982: Non-linear normal mode initialization with physics. Tech. Rep. No. 82-7, Dept. Meteorology, Florida State University, 57 pp.
65. Kreiss, H. O., 1979: Problems with different time scales for ordinary differential equations. SIAM. J. Num. Anal., 16, 980-998.
66. Kreiss, H. and J. Oliger, 1973: Methods for the approximate solution of time dependent problems. GARP Publ. Series, No. 10, WMO-ICSU, Geneva, Switzerland, 107 pp.
67. Lamb, H., 1932: Hydrodynamics, Dover Publications, New York, (6th Edition), 738 pp.
68. Leith, C. E., 1980: Non-linear normal mode initialization and quasi-geostrophic theory. J. Atmos. Sci., 37, 954-964.

69. Lindzen, R. S. and D. Blake, 1972: Lamb waves in the presence of realistic distributions of temperature and dissipation. J. Geophys. Res., 77, 2166-2176.

70. Longuet-Higgins, M.S., 1968: The eigenfunctions of Laplace's tidal equations over a sphere. Phil. Trans. Roy. Soc. London, A 262, 511-607.

71. Lorenc, A., 1981: A global three-dimensional multivariate statistical interpolation scheme. Mon. Weather Rev., 109, 701-721.

72. Lorenz, E. N., 1980: Attractor sets and quasi-geostrophic equilibrium. J. Atmos. Sci., 37, 1685-1699.

73. Lusternik, L. A. and V. J. Sobolev, 1961: Elements of functional analysis. Gordon and Breach, New York, 402 pp.

74. Machenhauer, B., 1977: On the dynamics of gravity oscillations in a shallow water model, with applications to normal mode initialization. Beitr. Phys. Atmos., 50, 253-271.

75. Machenhauer, B., 1979: The spectral method. GARP Publ. Series, No. 17, Vol. 2, WMO-ICSU, Geneva, Switzerland, 121-275.

76. Machenhauer, B. and R. Daley, 1972: A baroclinic primitive equation model with a spectral representation in three dimensions. Rep. No. 4, Inst. for Theoretical Meteor., Copenhagen University, Denmark, 62 pp.

77. Madala, R. V., 1981: Efficient time integration schemes for atmosphere and ocean models. Finite-difference Techniques for Vectorized Fluid Dynamics Calculations (D. Book, Ed.), Springer-Verlag, New York, 56-74.

78. Madden, R. A., 1979: Observations of large-scale traveling Rossby waves. Rev. Geophys. Space Phys., 17, 1935-1949.

79. Marchuk, G. I., 1974: Numerical methods in weather prediction, Academic Press, New York, 277 pp.

80. Margules, M., 1983: Luftbewegungen in einer rotierenden Sphäroidschale, Theil II. Trans. by B. Haurwitz, Air motions in a rotating spheroidal shell. NCAR Technical Note, 156+STR, National Center for Atmospheric Research, Boulder, CO.

81. Matsuno, T., 1966: Quasi-geostrophic motions in the equatorial area. J. Meteor. Soc. Japan, 44, 25-43.

82. McPherson, R.,1980: Lecture notes on operational analysis and assimilation of meteorological observations, Vols. 1-6, available from the author from National Meteorological Center, NOAA, Washington D.C.

83. Merilees, P. E., P. Ducharme and G. Jacques, 1977: Experiments with a polar filter and a one-dimensional semi-implicit algorithm. Atmosphere, 15, 19-32.

84. Mesinger, M. and A. Arakawa, 1976: Numerical methods used in atmospheric models: Vol. 1. GARP Publ. Series No. 17, WMO-ICSU, Geneva, Switzerland, 64 pp.

85. Miyakoda, K., R. Strickler and J. Chlodzinski, 1978: Initialization with the data assimilation method. Tellus, 30, 32-54.

86. Morse, P. M. and H. Feshbach, 1953: Methods of theoretical physics, Part II. McGraw-Hill, New York.

87. Orszag, S. A., 1970: Transform method for the calculation of vector-coupled sums: Application to the spectral form of the vorticity equation. J. Atmos. Sci., 27, 890-895.

88. Phillips, N. A., 1957: A coordinate system having some special advantages for numerical forecasting. J. Meteor., 14, 184-185.

89. Phillips, N. A., 1960: On the problem of initial data for the primitive equations. Tellus, 12, 121-126.

90. Phillips, N., 1981: Treatment of normal and abnormal modes. Mon. Weather Rev., 109, 1117-1119.

91. Phillips, N. A., 1982: On the completness of multivariate optimum interpolation for large-scale meteorological analysis. Mon. Weather Rev., 110, 1329-1334.

92. Platzman, G. W., 1962: The analytical dynamics of the spectral vorticity equation. J. Atmos. Sci., 19, 313-328.

93. Platzman, G. W., 1972: Two-dimensional free oscillations in natural basins. J. Phys. Ocean., 2, 117-138.

94. Platzman, G. W., 1979: The ENIAC computations of 1950 - Gateway to numerical weather prediction. Bull. Amer. Meteor. Soc., 60, 302-312.

95. Puri, K. and W. Bourke, 1982: A scheme to retain the Hadley circulation during nonlinear normal mode initialization. Mon. Weather Rev., 110, 327-335.
96. Robert, A., 1979: The semi-implicit method. GARP Publ. Series, No. 17, Vol. 2, WMO-ICSU, Geneva, Switzerland, 417-437.
97. Rossby, C. G. and Collaborators, 1939: Relation between variations in the intensity of the zonal circulation of the atmosphere and the displacements of the semi-permanent centers of action. J. Marine Res., 2, 38-55.
98. Salby, M., 1981: The 2-day wave in the middle atmosphere: Observations and theory. J. Geophys. Res., 86, 9654-9660.
99. Sasaki, Y., 1970: Some basic formalisms in numerical variational analysis. Mon. Weather Rev., 98, 875-883.
100. Schlatter, T. W., G. W. Branstator and L. Thiel, 1976: Testing a global multi-variate statistical objective analysis scheme with observed data. Mon. Weather Rev., 104, 765-783.
101. Sela, J. G., 1980: Spectral modeling at the National Meteorological Center. Mon. Weather Rev., 108, 1279-1292.
102. Shigehisa, Y., 1983: Normal modes of the shallow water equations for zonal wavenumber zero. Submitted to J. Meteor. Soc. Japan.
103. Somerville, R. C. J., 1980: Tropical influences on the predictability of ultralong waves. J. Atmos. Sci., 37, 1141-1156.
104. Staniforth, A. N. and R. W. Daley, 1977: A finite-element formulation for the vertical discretization of sigma-coordinate primitive equation models. Mon. Weather Rev., 105, 1108-1118.
105. Swarztauber, P. N. and A. Kasahara, 1983: The vector harmonic analysis of the Laplace tidal equations. To be submitted to SIAM J. Sci. Statist. Comp.
106. Talagrand, O., 1981: A study of the dynamics of four-dimensional data assimilation. Tellus, 33, 43-60.
107. Taylor, G. I., 1936: The oscillations of the atmosphere. Proc. Roy. Soc. London, A156, 318-326.

108. Temperton, C. and D. L. Williamson, 1981: Normal mode initialization for a multilevel grid-point model, Part I: Linear aspects. Mon. Weather Rev., 109, 729-743.

109. Thaning, L., 1983: On the existence of solutions to Machenhauer's nonlinear normal mode initialization. Submitted to Tellus.

110. Thompson, P. D., 1979: The mathematics of meteorology. Mathematics Today: Twelve informed essays. Springer-Verlag, New York, 127-152.

111. Tribbia, J. J., 1979: Nonlinear initialization on an equatorial beta-plane. Mon. Weather Rev., 107, 704-713.

112. Tribbia, J. J., 1981: Nonlinear normal mode balancing and the ellipticity condition. Mon. Weather Rev., 109, 1751-1761.

113. Tribbia, J. J., 1982: On variational normal mode initialization. Mon. Weather Rev., 110, 455-470.

114. Tribbia, J. J., 1983: A simple scheme for high order nonlinear normal mode initialization. To appear in Mon. Weather Rev.

115. Umscheid, L. and P. R. Bannon, 1977: A comparison of three global grids used in numerical prediction models. Mon. Weather Rev., 105, 618-635.

116. Wahba, G., 1982: Variational methods in simultaneous optimum interpolation and initialization. Proc. the (14th) Stanstead Seminar, NCAR/TN-204+PROC, 178-185.

117. Wahba, G. and J. Wendelberger, 1980: Some new mathematical methods for variational objective analysis using splines and cross validation. Mon. Weather Rev., 108, 1122-1143.

118. Wallace, J. M., S. Tibaldi and A. J. Simmons, 1983: Reduction of systematic forecast errors in the ECMWF model through the introduction of an envelope orography. To appear in Quart. J. Roy. Meteor. Soc.

119. Williamson, D. L., 1976: Normal mode initialization procedure applied to forecasts with the global shallow water equations. Mon. Weather Rev., 104, 195-206.

120. Williamson, D. L., 1979: Difference approximations for fluid flow on a sphere. GARP Publ. Series, No. 17, Vol. 2, WMO-ICSU, Geneva, Switzerland, 51-120.

121. Williamson, D. L., 1983: Description of NCAR Community Climate Model (CCMOB). NCAR/TN-210+STR, 88 pp.

122. Williamson, D. L. and R. Daley, 1983: A unified analysis-initialization technique. To appear in Mon. Weather Rev.

123. Williamson, D. L. and C. Temperton, 1981: Normal mode initialization for a multilevel grid-point model, Part II: Nonlinear aspects. Mon. Weather Rev., 109, 744-757.

National Center for Atmospheric Research
P. O. Box 3000
Boulder, Colorado 80307

SYSTOLIC ALGORITHMS

H. T. Kung

1. INTRODUCTION

Much of the time consuming "front-end processing," that deals with large amounts of data obtained directly from sensors in application areas such as signal and image processing, is inherently regular and parallel. By exploiting this regularity and parallelism systolic architectures, when properly implemented, can achieve high performance with low cost.

The basic principle of a systolic architecture, a systolic array in particular, is illustrated in Figure 1. By replacing a single processing element (PE) with an array of PEs, a higher computation throughput can be achieved without increasing memory bandwidth. The function of the memory in the diagram is analogous to that of the heart; it "pulses" data (instead of blood) through the array of PEs. The crux of this approach is to ensure that once a data item is brought out from the memory it can be used effectively at each PE it passes while being "pumped" from PE to PE along the array. As we will see later in this section, recent research has shown that systolic architectures are possible for a wide class of compute-bound computations where multiple operations are performed on each data item in a repetitive manner.

Being able to use each input data item a number of times is just one of the many advantages of a systolic architecture. Other advantages, including modular expansability, simple and regular data and control flows, use of simple and uniform cells, and elimination of global broadcasting and fan-in, are also characteristic [17]. These properties are highly desirable for VLSI implementations; indeed the

ISBN 0-12-546080-5

Instead of:

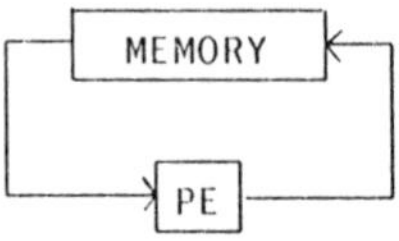

we have:

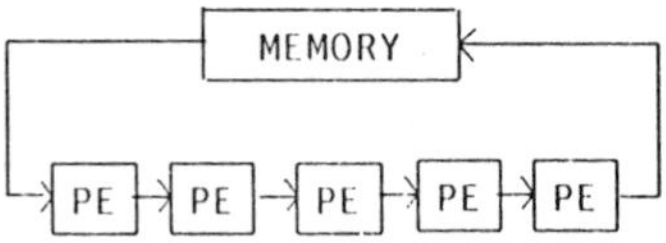

Figure 1. Basic principle of a systolic array.

advances in VLSI technology have been a major motivation for the systolic architecture research.

Mathematical algorithms implementable with systolic architectures are called systolic algorithms. A systolic algorithm has many systolic cells, which can operate in parallel, and each of them is to be implemented by a PE in a systolic architecture. A large number of systolic algorithms are known today and several theoretical frameworks for the design of systolic algorithms are being developed [20, 24, 31]. Extensive references for the systolic literature can be found in separate papers [7, 17]. To indicate the breadth of the systolic approach, the following is a partial list of problems for which systolic solutions exist:

- Signal and image processing:
  - FIR and IIR filtering and 1-D convolution;
  - 2-D convolution and correlation;
  - discrete Fourier transform;
  - interpolation;
  - 1-D and 2-D median filtering; and
  - geometric warping.

- Matrix arithmetic:
  - matrix-vector multiplication;
  - matrix-matrix multiplication;
  - matrix triangularization (solution of linear systems, matrix inversion);
  - solution of triangular linear systems;
  - solution of Toeplitz linear systems;
  - QR-decomposition (least squares computations, covariance matrix inversion);
  - singular value decomposition; and
  - eigenvalue problems.

- Non-numeric applications:
  - o data structures--stack and queue, priority queue, searching, and sorting;
  - o graph algorithms--transitive closure, minimum spanning trees, and connected components;
  - o geometric algorithms--convex hull generation;
  - o language recognition--string matching and regular expression;
  - o dynamic programming;
  - o polynomial manipulation--multiplication, division, and greatest common divisor;
  - o integer greatest common divisors; and
  - o relational data-base operation.

In the next section we review an important systolic algorithm, presented in [12], for matrix triangularization and discuss one of its applications in adaptive signal processing. Designing systolic algorithms, however, just constitutes the very beginning phase of the total effort of constructing a systolic device for real-life application use. Activities following this phase include at least architectural specification of the systolic processor, processor implementation, integration to the host system and application software development. In Sections 3 and 4 we examine some of the issues related to the implementation of systolic algorithms.

2. MATRIX TRIANGULARIZATION: A SYSTOLIC ALGORITHM EXAMPLE

Given an $n \times p$ Hermitian matrix X with $p \le n$, the triangularization problem is to determine an $n \times n$ "multiplier matrix" M such that

$$MX = \begin{bmatrix} U \\ 0 \end{bmatrix}$$

where U is a $p \times p$ upper triangular matrix, and furthermore to compute the entries in U. By triangularization, many problems in matrix computation can be reduced to that of solving triangular linear systems. This is the major step in all direct methods for solving linear systems. When M is restricted to be an orthogonal matrix Q, it is also the key step in computing least squares solutions and in the QR algorithm for computing eigenvalues. A triangular systolic array is ideal for performing orthogonal triangularization. Figure 2 depicts the systolic array for the case when $p=5$. The array consists of two types of cells, internal cells (represented by squares) and boundary cells (represented by circles). Internal cells basically perform multiplies and adds, whereas boundary cells perform divisions or reciprocals plus possibly other operations. During the computation, X enters the systolic array from its top boundary one row at a time in a skewed order. Current entries in the i-th row of U are kept in the i-th row of the systolic array, one entry at each cell. Initially zeros are stored in all cells, and when computation is complete, entries in U will be readily read out, one from each cell.

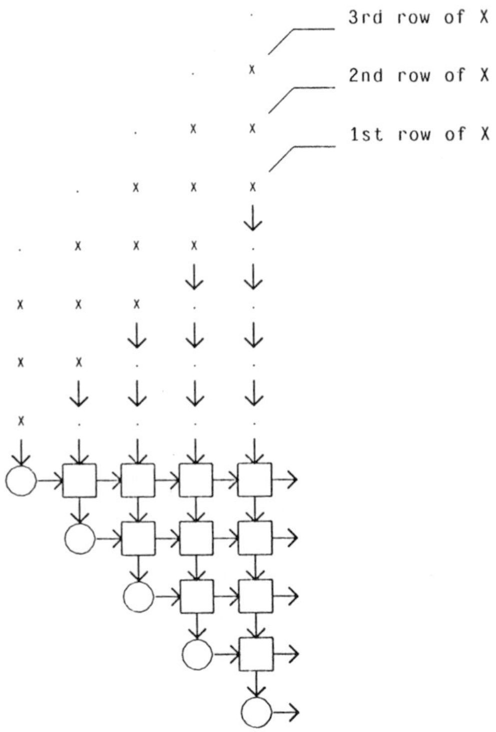

Figure 2. Triangular systolic array for matrix triangularization.

The top row of the systolic array turns every arriving row of X into a row with its first entry being zero, and outputs results to the second row of the systolic array. Similarly, the second row of the systolic array turns every row of X it receives from the row above into a row with its second entry being zero, and outputs results to the third row of the systolic array, and so on. While updating a row of X, a row of the systolic array also updates current entries of U that are stored in its cells. The boundary cell at the left end determines the "plane rotation" parameters needed for both updates, and they are sent to the right to be used in actual updatings taking place at internal cells. Because rows of X pass through the systolic array in the skewed order, parameters determined at boundary cells will reach internal cells at right times.

We have described the triangular systolic array for computing U. An important fact to realize is that U is computed on-the-fly, in the sense that rows of X can keep arriving at the top boundary of the systolic array, while the entries of U for earlier rows of X are being computed inside the systolic array. This property allows the use of the systolic array to perform the rank-one updating of the Cholesky decomposition of a time-varying covariance matrix, a crucial step in adaptive signal processing [2, 25]. In this case each row x of X represents measurements from a sensor array in a time instant. For each newly arriving sample x the covariance matrix R is updated by:

$$R_{new} = \mu R_{old} + (1-\mu) x x^+,$$

where x^+ is the conjugate transpose of x and μ is an arbitrary but fixed weighting factor between 0 and 1. Suppose that the Cholesky decomposition of R_{old} is known, that is, we know an $n \times n$ upper triangular matrix U_{old} such that

$$R_{old} = U_{old}^+ U_{old}.$$

We would like to find a new $n \times n$ upper triangular matrix U_{new} such that

$$R_{new} = U_{new}^+ U_{new}.$$

It turns out that U_{new} can be computed directly from U_{old} without even having to form R_{new}. To see this, consider the $(n+1) \times n$ matrix obtained by adding the weighted new data, $(1-\mu)^{1/2} x^+$, as a row to the top of the weighted old $n \times n$ upper triangular matrix $\mu^{1/2} U_{old}$, that is,

$$\bar{U} = \begin{bmatrix} (1-\mu)^{1/2} x^+ \\ \mu^{1/2} U_{old} \end{bmatrix}$$

This new $(n+1) \times n$ matrix $\bar{U}$ is a Hessenberg matrix. In order to continue the computation for the next time sample, we reduce $\bar{U}$ to an $n \times n$ upper triangular matrix. This may be accomplished by the triangular systolic array described in the beginning of this section. More precisely, by orthogonal triangularization of $\bar{U}$ via the systolic array, we obtain an $n \times n$ upper triangular matrix U_{new} such that

$$Q\bar{U} = \begin{bmatrix} U_{new} \\ 0 \end{bmatrix}$$

By the following expression, we see that $U_{new}^+ U_{new}$ is the desired Cholesky decomposition of R_{new}:

$$R_{new} = \bar{U}^+ \bar{U} = \bar{U}^+ Q^+ Q\bar{U} = (Q\bar{U})^+ (Q\bar{U}) = U_{new}^+ U_{new}.$$

Therefore the Cholesky decomposition can be updated on-the-fly by using the systolic array to perform the orthogonal triangularization. The "weighting" factors μ and $1-\mu$ can be easily incorporated into the system by having the boundary cells generate proper weights. From the Cholesky decomposition adaptive weights are readily computed by solving triangular linear systems, which in turn can be carried out by a linear systolic array [19]. We see that the systolic array for matrix triangularization together with a systolic array for solving triangular linear systems can potentially compute adaptive weights in real-time, a task that is computationally intractable today. For certain static signal environment, R is Toeplitz. In this case only a one-dimensional linear systolic array, rather than the two-dimensional, triangular systolic array, is needed [4].

3. IMPLEMENTATION ALTERNATIVES

Many alternatives exist for the implementation of systolic algorithms at both chip and board levels. We discuss them along two dimensions: flexibility and interconnection topology.

3.1. Flexibility

1. Single-purpose systolic array [6, 11, 22, 28, 30]. The most straightforward way to implement a systolic algorithm is to build a special-purpose systolic array processor just for that algorithm. This approach is reasonable if one or more of the following conditions hold: (1) the performance of the processor is of ultimate importance and the use of the processor is well-understood; (2) the processor will be used in large quantities despite the fact that it is single-purpose; and (3) the design and implementation cost of the processor is low--this is the case for those systolic arrays consisting of only few types of very simple cells like some of the pattern matching and correlation arrays.

2. Multi-purpose systolic array [1, 32]. This type of systolic processors can implement a predefined set of systolic algorithms. The approach is based on the observation that many systolic algorithms like those for convolution and matrix multiply can be executed on systolic arrays of very similar structures. Therefore a systolic processor, with little overheads for providing the necessary flexibility, can perform a number of functions under some simple controls.

3. Non-programmable building-block [16]. Some common functions such as multiplier-accumulation are performed by cells of a large number of systolic arrays. Thus it is possible to construct building-block processors capable of executing a few predefined, commonly used functions, and then connect them to form a variety of systolic processors of different sizes and shapes. A good building-block should meet the needs from current as well as future systolic algorithms.

4. Programmable building-block [9, 10, 26]. In this approach building-blocks, by programming, can implement a large family

of systolic cells. Of course the programmable approach is not as efficient as the non-programmable one, because of the overheads for supporting the programmability. Nevertheless it fulfills the need of implementing those systolic arrays, which each are not important enough for warranting individual, custom hardware implementation, but in aggregate can be implemented cost-effectively by some programmable building-blocks. Moreover in some systolic designs, the instruction that the cell executes in a cycle depends, in a complicated way, on the inputs to the cell and the cell's state during that cycle. A notable example is the the systolic array for computing greatest common divisors [3]. The programmable approach seems to be the only effective way to implement those complicated, data-dependent systolic arrays.

5. Programmable systolic arrays [6, 27]. These are systolic array processors where a fixed number of programmable PEs are connected together in certain manner, possibly with other control circuits. In particular when a number of programmable building-blocks mentioned above are connected into a fixed array, they form such a programmable systolic array. But it is possible that a programmable systolic array is assembled from PEs which are not intended to be programmable building-blocks for any other systolic arrays. Programmable systolic arrays are more flexible than the multi-purpose systolic arrays in the sense that the PEs are programmable and sometimes even their interconnections can be configured by software control before a computation starts.

3.2. Interconnection Topology

Systolic arrays typically call for simple and regular array interconnections among their PEs. We classify interconnections for n-PE arrays into three classes according to the number of computations performed for each I/O operation:

1. 2-D (two-dimensional) systolic arrays, such as square and hexagonal arrays, for which at each cycle, $O(n)$ PEs perform computations while $O(\sqrt{n})$ boundary PEs perform I/O, and thus the computation over I/O ratio is $O(\sqrt{n})$. Examples are systolic arrays for matrix operations [12, 19].

2. Degenerated 2-D systolic arrays, that is, 2-D arrays consisting of only one row of PEs, for which at each cycle, $O(n)$ PEs perform computations and $O(n)$ PEs perform I/O, and thus the computation over I/O ratio is $O(1)$. Examples are systolic arrays for solution of triangular linear systems [19] and orthogonal transformations [14].

3. Linear or 1-D systolic arrays, for which at each cycle, $O(n)$ PEs perform computations but only the two end PEs perform I/O, and thus the computation over I/O ratio is $O(n)$. Examples are various systolic convolution arrays [17].

For situations where the communication bandwidth between the host system and the systolic array is a bottleneck, linear arrays are preferable to 2-D arrays and degenerated 2-D arrays are most undesirable. Sometimes for a given problem there could be both linear and 2-D systolic array solutions. For example, two-dimensional convolution can be performed by a linear systolic array [21, 18] or by a 2-D systolic array [22]. In this case we can choose a linear solution to solve the communication bandwidth problem. Linear arrays (and degenerated 2-D arrays) have the additional advantage that they can always be safely synchronized by a global clock [8]. To avoid the excessive I/O bandwidth required by degenerated 2-D arrays, it may sometimes be cost-effective to associate each PE with a memory, where data required by the PE can be cached [1, 32].

In general some "design-time" rather than "run-time" flexibility at the interconnection level should always be provided not only for implementing different interconnections, but also for system maintenance, debugging and fault-tolerance. The idea is that by modifying interconnections faulty PEs can be detected, and replaced or by-passed during a computation. The usefulness of run-time reconfigurability for systolic array processors is not clear, since interconnections typically do not change during a computation.

4. ALGORITHMS REFLECTING IMPLEMENTATION NEEDS

In spite of the large number of systolic algorithms already designed, a major effort is still needed in developing systolic algorithms that are convenient for implementation. Currently, when implementing a systolic algorithm on a systolic processor, we often find that the algorithm must be substantially modified and in some cases even totally new algorithms have to be invented. It is useful to know the following experiences:

- <u>Systolic algorithms using linear arrays</u>. For implementation convenience, systolic algorithms using linear arrays are superior than those using 2-D arrays--it is not an accident that most of the systolic algorithms that have been implemented use linear arrays. Besides for reasons noted earlier that linear systolic arrays enjoy the advantages of low I/O bandwidth requirements and ease of synchronization by clocking, we find it is convenient to balance the computation over I/O ratio for algorithms using linear arrays. As illustrated in Figure 3, the ratio increases linearly with the number of systolic cells allocated to each PE. For example, suppose that in a systolic algorithm the boundary cell (denoted by the circle in the figure) takes k times more operations than the internal cells (denoted by squares). In implementation we can achieve a balancing in speed between the two types of cells by simply allocating k rather than one internal cell to each PE.

- <u>Systolic algorithms for pipelined units</u>. Systolic array processors often employ PEs which are pipelined units themselves. This is typically true for floating-point processing elements, which could have four to ten pipeline stages. For

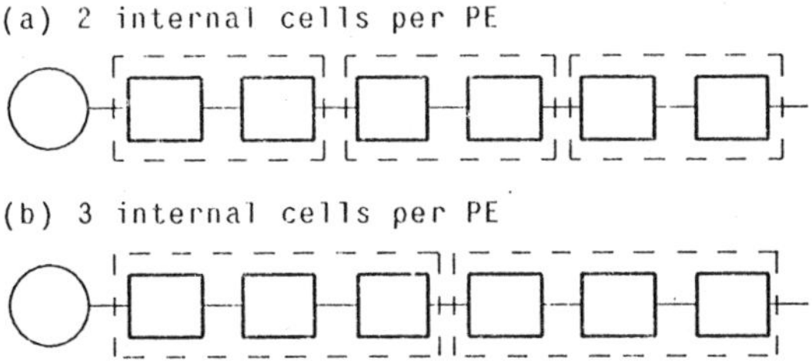

Figure 3. Balancing I/O and computation for linear array:
(a) 3 cell operations per I/O; (b) 2 cell operations per I/O.

performance, it is important that one can take advantage of this second level of pipelining in a systolic implementation [18].

- Problem decomposition. In practice, computational problems are often larger than what a given systolic array can handle in one pass. In this case an original problem must be decomposed into subproblems, and the systolic array then processes these subproblems one at a time. Some systolic algorithms can be easily modified to work on subproblems with relatively small I/O and control overheads, but some can not [15, 23]. The design of a systolic algorithm is really not complete unless its decomposition methods are provided, although partitioning schemes are sometimes self-evident once the basic algorithm is presented.

- Systolic algorithms without cycles. Linear systolic arrays where data all flow in one direction are examples of systolic algorithms without cycles, i.e., without feedback loops. Systolic arrays that use bi-directional data flows are examples of systolic algorithms with cycles. We know that certain problems such as convolution, graph connectivity, and graph transitive closure can be solved by systolic algorithms either with or without cycles [13, 17, 29]. It turns out that systolic designs without cycles are easier to use than ones with cycles. For example, by slight modification the second level of pipelining mentioned above can always be achieved for systolic algorithms without cycles, but without decreasing throughput this is impossible for algorithms with cycles. Moreover, it is easy to deal with problem decomposition and to achieve fault tolerance for systolic designs without cycles.

Being simple and regular is not everything about systolic algorithm design; for implementation convenience, other issues such as balancing computation with I/O, use of pipelined processing elements, problem decomposition and fault tolerance are crucial.

5. SUMMARY

In this paper we reviewed a systolic algorithm for matrix triangularization, and illustrated one of its important uses in the on-the-fly updating of Cholesky decomposition of covariance matrices--a crucial computation in adaptive signal processing. We examined issues and alternatives for the implementation of systolic algorithms in general, and provided some guidelines for designing systolic algorithms that will be convenient for implementation. As stated many systolic algorithms have been designed in recent years; major efforts now started in building systolic processors and in using them for large, real-life applications. Practical issues on the implementation and use of systolic array processors in systems are beginning to receive substantial attention as well. Readers can follow some of the references of this paper for further information on systolic algorithms and their implementation.

References

[1] Blackmer, J., Frank, G. and Kuekes, P.
A 200 Million Operations per Second (MOPS) Systolic Processor.
In *Proceedings of SPIE Symposium, Vol. 298, Real-Time Signal Processing IV*, pages 10-18. Society of Photo-Optical Instrumentation Engineers, August, 1981.

[2] Bowen, B. A. and Brown, W. R.
VLSI Systems Design for Digital Signal Processing. Volume 1: *Signal Processing and Signal Processors*.
Prentice-Hall, Inc., Englewood Cliffs, New Jersey, 1982.

[3] Brent, R.P. and Kung, H.T.
Systolic VLSI Arrays for Polynomial GCD Computation.
Technical Report, Carnegie-Mellon University, Computer Science Department, May, 1982.
The paper is to appear in *IEEE Transactions on Computers* 1983.

[4] Brent, R.P. and Luk, F.T.
A Systolic Array for the Linear-Time Solution of Toeplitz Systems of Equations.
Journal of VLSI and Computer Systems 1(1):1-22, 1983.

[5] Bromley, K., Symanski, J.J., Speiser, J.M., and Whitehouse, H.J.
Systolic Array Processor Developments.
In Kung, H.T., Sproull, R.F., and Steele, G.L., Jr. (editors), *VLSI Systems and Computations*, pages 273-284. Computer Science Department, Carnegie-Mellon University, Computer Science Press, Inc., October, 1981.

[6] Corry, A. and Patel, K.
A CMOS/SOS VLSI Correlator.
In *Proceedings of 1983 International Symposium on VLSI Technology, Systems and Applications*, pages 134-137. 1983.

[7] Fisher, A.L. and Kung, H.T.
Special-Purpose VLSI Architectures: General Discussions and a Case Study.
In *Proceedings of the USC Workshop on VLSI and Modern Signal Processing*. November, 1982.

[8] Fisher, A.L. and Kung, H.T.
Synchronizing Large VLSI Processor Arrays.
In *Proceedings of the 10th Annual International Symposium on Computer Architecture*, pages 54-58. June, 1983.

[9] Fisher, A.L., Kung, H.T., Monier, L.M. and Dohi, Y.
Architecture of the PSC: A Programmable Systolic Chip.
In *Proceedings of the 10th Annual International Symposium on Computer Architecture*, pages 48-53. June, 1983.

[10] Fisher, A.L., Kung, H.T., Monier, L.M., Walker, H. and Dohi, Y.
Design of the PSC: A Programmable Systolic Chip.
In Bryant, R. (editor), *Proceedings of the Third Caltech Conference on Very Large Scale Integration*, pages 287-302. California Institute of Technology, Computer Science Press, Inc., March, 1983.

[11] Foster, M.J. and Kung, H.T.
The Design of Special-Purpose VLSI Chips.
Computer Magazine 13(1):26-40, January, 1980.
Reprint of the paper appears in *Digital MOS Integrated Circuits*, edited by Elmasry, M.I., IEEE Press Selected Reprint Series, 1981, pp. 204-217.

[12] Gentleman, W.M. and Kung, H.T.
Matrix Triangularization by Systolic Arrays.
In *Proceedings of SPIE Symposium, Vol. 298, Real-Time Signal Processing IV*, pages 19-26. Society of Photo-Optical Instrumentation Engineers, August, 1981.

[13] Guibas, L.J., Kung, H.T. and Thompson, C.D.
Direct VLSI Implementation of Combinatorial Algorithms.
In *Proceedings of Conference on Very Large Scale Integration: Architecture, Design, Fabrication*, pages 509-525. California Institute of Technology, January, 1979.

[14] Heller, D.E. and Ipsen, I.C.F.
Systolic Networks for Orthogonal Equivalence Transformations and Their Applications.
In *Proceedings of Conference on Advanced Research in VLSI*, pages 113-122. Massachusetts Institute of Technology, Cambridge, Massachusetts, January, 1982.

[15] Hong, J.-W. and Kung, H.T.
I/O Complexity: The Red-Blue Pebble Game.
In *Proceedings of the Thirteenth Annual ACM Symposium on Theory of Computing*, pages 326-333. ACM SIGACT, May, 1981.
To appear in *Journal of the ACM*.

[16] Kuekes, P. and Shen, J.
One-Gigaflop VLSI Systolic Processor.
In *Proceedings of SPIE Symposium, Vol. 431, Real-Time Signal Processing VI*. Society of Photo-Optical Instrumentation Engineers, August, 1983.

[17] Kung, H.T.
Why Systolic Architectures?
Computer Magazine 15(1):37-46, January, 1982.

[18] Kung, H.T., Ruane, L.M., and Yen, D.W.L.
Two-Level Pipelined Systolic Array for Multidimensional Convolution.
Image and Vision Computing 1(1):30-36, February, 1983.
A preliminary version appears in *VLSI Systems and Computations*, edited by H. T. Kung, G. L. Steele, Jr. and R. F. Sproull, Computer Science Press, Maryland, 1981, pp. 255-264.

[19] Kung, H.T. and Leiserson, C.E.
Systolic Arrays (for VLSI).
In Duff, I. S. and Stewart, G. W. (editors), *Sparse Matrix Proceedings 1978*, pages 256-282. Society for Industrial and Applied Mathematics, 1979.
A slightly different version appears in *Introduction to VLSI Systems* by C. A. Mead and L. A. Conway, Addison-Wesley, 1980, Section 8.3.

[20] Kung, H.T. and Lin, W.T.
An Algebra for Systolic Computation.
In *Proceedings of Conference on Elliptic Problem Solvers*. Academic Press, January, 1983.

[21] Kung, H.T. and Picard, R.L.
Hardware Pipelines for Multi-Dimensional Convolution and Resampling.
In *Proceedings of the 1981 IEEE Computer Society Workshop on Computer Architecture for Pattern Analysis and Image Database Management*, pages 273-278. IEEE Computer Society Press, November, 1981.

[22] Kung, H.T. and Song, S.W.
A Systolic 2-D Convolution Chip.
In Preston, K., Jr. and Uhr, L. (editor), *Multicomputers and Image Processing: Algorithms and Programs*, pages 373-384. Academic Press, 1982.
An extended abstract appears in *Proceedings of 1981 IEEE Computer Society Workshop on Computer Architecture for Pattern Analysis and Image Database Management*, November 11-13, 1981, pp. 159-160.

[23] Kung, H.T. and Yu, S.Q.
Integrating High-Performance Special-Purpose Devices into a System.
In Randel, B. and Treleaven, P.C. (editors), *VLSI Architecture*, pages 205-211. Prentice/Hall International, 1983.

[24] Leiserson, C.E. and Saxe, J.B.
Optimizing Synchronous Systems.
Journal of VLSI and Computer Systems 1(1):41-68, 1983.

[25] Monzingo, R. A. and Miller, T. W.
Introduction to Adaptive Arrays.
John Wiley & Sons, Inc., New York, 1980.

[26] Sorasen, O. and Solberg, B.
VLSI Implemented Systolic Array Processor.
In *Proceedings of VLSI 83 International Conference*. Trondheim, Norway, August, 1983.

[27] Symanski, J.J.
Systolic Array Processor Implementation.
In *Proceedings of SPIE Symposium, Vol. 298, Real-Time Signal Processing IV*, pages 27-32. Society of Photo-Optical Instrumentation, August, 1981.

[28] Tamura, P.N., Haugen, P.R. and Betz, B.K.
Time Integrating Digital Correlator.
In *Proceedings of SPIE Symposium, Vol. 431, Real-Time Signal Processing VI*. Society of Photo-Optical Instrumentation Engineers, August, 1983.

[29] Tchuente, M. and Melkemi, L.
Systolic Arrays for Connectivity Problems and Triangularization for Band Matrices.
Technical Report R.R. No. 366, IMAG, Institut National Polytechnique de Grenoble, March, 1983.

[30] Travassos, R.H.
A Systolic Signal Processor for Recursive Filtering.
Integrated Systems, Inc., Palo Alto, California.
December, 1982.

[31] Weiser, U. and Davis, A.
A Wavefront Notation Tool for VLSI Array Design.
In Kung, H.T., Sproull, R.F., and Steele, G.L., Jr. (editors), *VLSI Systems and Computations*, pages 226-234. Computer Science Department, Carnegie-Mellon University, Computer Science Press, Inc., October, 1981.

[32] Yen, D.W.L. and Kulkarni, A.V.
Systolic Processing and an Implementation for Signal and Image Processing.
IEEE Transactions on Computers C-31(10):1000-1009, October, 1982.

This research was supported in part by the Office of Naval Research under Contracts N00014-76-C-0370, NR 044-422 and N00014-80-C-0236, NR 048-659.

Department of Computer Science
Carnegie-Mellon University
Pittsburgh, PA 15213

AN APPROACH TO FLUID MECHANICS CALCULATIONS ON SERIAL AND PARALLEL COMPUTER ARCHITECTURES

Dennis R. Liles, John H. Mahaffy, and Paul T. Giguere

1. SUMMARY

The Transient Reactor Analysis Code (TRAC) is a large FORTRAN thermal-hydraulics program designed to solve problems involving internal flows in nuclear reactors. The current versions have been designed for a CDC 7600 and CRAY 1 but benchmarks have been run in parallel simulations. This paper will discuss the methods in use, the reason that these techniques are effective, and their extension to parallel machines.

2. INTRODUCTION

The Transient Reactor Analysis Code (TRAC) is a reasonably large (over 50000 FORTRAN lines) thermal-hydraulic computer program designed to analyze nuclear reactor transients. The bulk of the fluids representation is performed by an implicit hydrodynamics package in one spatial dimension. A three-dimensional semi-implicit model is available to model the reactor vessel. One- and two-dimensional conduction solutions are used to account for the stored and generated thermal energy in the walls and fuel rods. TRAC was designed to be run on a CDC 7600, although most of the large problems are now computed on a CRAY. Because of the wide variety of problems of interest, input descriptions may have as few as 30-40 mesh cells or as many

ISBN 0-12-546080-5

as 1500. The transients to be computed can be as short as 1/2 s or as long as several hours. Although a reasonable one-dimensional representation of the Three Mile Island accident can be executed 16 times faster than real time on a CRAY, some very complicated plant descriptions may run for many hours on the same machine.

We shall briefly discuss the equation set and the numerical techniques used to solve these equations.[1] We shall then give some approximate timing statistics on scalar machines and some comments on the degree to which the system can be vectorized. Finally, we will provide some information on the application of the methods to parallel computers.

3. THE MODELS

The most costly portion of the calculation normally is the hydrodynamic computation. A multicomponent representation is used to model two full fields (vapor and liquid). In addition, extra mass equations have been added to account for the presence of a noncondensable gas and to track a single material dissolved in the liquid. Boron is the normal solute and is injected into the primary system under certain conditions to act as a neutronic poison. The equations are:

Liquid Mass Equation

$$\frac{\partial(1-\alpha)\rho_\ell}{\partial t} + \nabla \cdot \left[(1-\alpha)\,\rho_\ell \vec{V}_\ell\right] = -\,\Gamma \quad . \tag{1}$$

Combined Vapor Mass Equation

$$\frac{\partial(\alpha\rho_g)}{\partial t} + \nabla \cdot (\alpha\rho_g\vec{V}_g) = \Gamma \quad . \tag{2}$$

Noncondensable Gas Mass Equation

$$\frac{\partial(\alpha\rho_a)}{\partial t} + \nabla \cdot (\alpha\rho_a\vec{V}_g) = 0 \quad . \tag{3}$$

Solute Concentration Equation

$$\frac{\partial(1-\alpha)m\,\rho_\ell}{\partial t} + \nabla \cdot [(1-\alpha)m\rho_\ell\vec{V}_\ell] = S_c \tag{3A}$$

Combined Vapor Equation of Motion

$$\frac{\partial\vec{V}_g}{\partial t} + \vec{V}_g \cdot \nabla\,\vec{V}_g = -\frac{1}{\rho_g}\nabla p - \frac{c_i}{\alpha\rho_g}(\vec{V}_g - \vec{V}_\ell)\,|\vec{V}_g - \vec{V}_\ell|$$

$$- \frac{\Gamma^+}{\alpha\rho_g}(\vec{V}_g - \vec{V}_\ell)$$

$$- \frac{c_{wg}}{\alpha\rho_g}\vec{V}_g\,|\vec{V}_g| + \vec{g} \quad . \tag{4}$$

Liquid Equation of Motion

$$\frac{\partial\vec{V}_\ell}{\partial t} + \vec{V}_\ell \cdot \nabla\,\vec{V}_\ell = -\frac{1}{\rho_\ell}\nabla p + \frac{c_i}{(1-\alpha)\rho_\ell}(\vec{V}_g - \vec{V}_\ell)\,|\vec{V}_g - \vec{V}_\ell|$$

$$- \frac{\Gamma^-}{(1-\alpha)\rho_\ell}(\vec{V}_g - \vec{V}_\ell)$$

$$- \frac{c_{w\ell}}{(1-\alpha)\rho_\ell}\vec{V}_\ell\,|\vec{V}_\ell| + \vec{g} \quad . \tag{5}$$

Combined Vapor Energy Equation

$$\frac{\partial}{\partial t}(\alpha\rho_g e_g) + \nabla \cdot (\alpha\rho_g e_g \vec{V}_g)$$

$$= -p\frac{\partial\alpha}{\partial t} - p\nabla \cdot (\alpha\vec{V}_g) + q_{wg} + q_{ig}$$

$$+ \Gamma h_{sg} \quad . \tag{6}$$

Total Energy Equation

$$\frac{\partial[(1-\alpha)\rho_\ell e_\ell + \alpha\rho_g e_g]}{\partial t} + \nabla \cdot [(1-\alpha)\rho_\ell e_\ell \vec{V}_\ell + \alpha\rho_g e_g \vec{V}_g]$$

$$= -p\nabla \cdot [(1-\alpha)\vec{V}_\ell + \alpha\vec{V}_g] + q_{w\ell}$$

$$+ q_{wg} \quad . \tag{7}$$

In these equations ρ is a microscopic density, e an internal energy, V a velocity, α the volume fraction of the vapor or gas, p is the static pressure, and m is the solute concentration. The subscripts ℓ and g refer to liquid and gas, respectively.

In these equations the vapor densities and energies are sums of the steam and noncondensable components,

$$\rho_g = \rho_s + \rho_a \tag{8}$$

and

$$\rho_g e_g = \rho_s e_s + \rho_a e_a \quad . \tag{9}$$

We assume Dalton's law applies; therefore,

$$p = p_s + p_a \quad . \tag{10}$$

A subscript, a, is used for the noncondensable gas because the internal thermodynamic properties model air. It would be easy to replace these properties with others describing different noncondensable gases.

In addition to the thermodynamic relations that are required for closure [Eq. (17)], specifications for the interfacial drag coefficients (c_i), the interfacial heat transfer (q_{ig}), the phase-change rate (Γ), the solute source term (S_c), the wall shear coefficients (c_{wg} and $c_{w\ell}$), and the wall heat transfers (q_{wg} and $q_{w\ell}$) are required. Gamma is evaluated from a simple thermal energy jump relation,

$$\Gamma = \frac{-q_{ig} - q_{i\ell}}{h_{sg} - h_{s\ell}} , \tag{11}$$

where

$$q_{ig} = h_{ig} A_i \frac{(T_{ss} - T_g)}{vol} \tag{12}$$

and

$$q_{i\ell} = h_{i\ell} A_i \frac{(T_{ss} - T_\ell)}{vol} . \tag{13}$$

Here A_i and the h_i terms are the interfacial area and heat-transfer coefficients and T_{ss} is the saturation temperature corresponding to the partial steam pressure. The term Γ^+ is equal to Γ for positive Γ and zero for negative Γ, and Γ^- is Γ when negative, otherwise zero. The quantities h_{sg} and $h_{s\ell}$ are the saturation enthalpies of the vapor and liquid, respectively.

Wall heat-transfer terms assume the form

$$q_{wg} = h_{wg} A_{wg} (T_w - T_g) \tag{14}$$

and

$$q_{w\ell} = h_{w\ell} A_{w\ell} \frac{(T_w - T_\ell)}{vol} \quad , \tag{15}$$

where A_{wg} and $A_{w\ell}$ are the heated surface areas of the cell.

An additional partial differential equation of the form

$$\rho_w C_{P_w} \frac{\partial T_w}{\partial t} = - \nabla \cdot K \nabla T_w + \dot{q}''' \tag{16}$$

is solved for the wall temperatures. The internal heat-generation term ($\dot{q}'''$) can be used to represent either the fission heat release or the ohmic heating characteristic of most experiments. Equation (16) is solved in one dimension in all regions except the outer thin shell (cladding) of the fuel rods where a two-dimensional (r - z) model is employed.

4. THE NUMERICS

A staggered first-order finite-difference scheme is used to approximate Eqs. (1-7). The basic solution technique involves linearizing the scalar equations and reducing them to a system in V_ℓ, V_g, T_g, T_ℓ, α, and P. This is accomplished by using the thermal equations of state,

$$\rho_\ell = \rho_\ell(P, T_\ell) \tag{17A}$$

and

$$\rho_g = \rho_g(P, T_g) \quad ; \tag{17B}$$

the caloric equations of state,

$$e_\ell = e_\ell(P, T_\ell) \tag{17C}$$

and

$$e_g = e_g(P, T_g) \quad ; \tag{17D}$$

and the definitions for ρ_m and e_m.

A further system reduction is accomplished by observing that the finite-difference vapor and liquid equations of motion can yield equations of the form

$$V^{n+1} = V^n + (\text{conv}^n + \frac{1}{\rho_\ell} \nabla P^{n+1} + \text{Fric}) \Delta t \qquad (18)$$

if the $V\nabla V$ (designated conv) terms are treated explicitly and the ∇P terms are implicit. Fric is both the interfacial and wall friction terms and is treated partially implicitly; n and n + 1 refer to the old and new time levels. This produces a semi-implicit scheme that imposes a stability limit of the form

$$V \frac{\Delta t}{\Delta X} < 1 \quad , \qquad (19)$$

where V is a material velocity. The very high sound speeds in the liquid do not impose an additional stability limit. Using vapor and liquid equations of the form of Eq. (18), the system of variables may be reduced further to T_ℓ, T_g, p, and α and can then be solved.

An improvement to the method proposed in Ref. 2 has been implemented to reduce the computing cost for the multi-dimensional vessel component. The linear system that results from this method is a block seven-stripe matrix in the three-dimensional component. In performing the Gauss-Seidel operation, if the nonlinear terms are not updated, the matrix coefficients remain constant for the time step. In this case a Gauss elimination technique can be applied once at each time step to the seven-stripe block array that allows its reduction to a seven-stripe single-element array. This results in a much faster iteration (after the first iteration) for the pressure. When the vessel pressures are obtained to a specified convergence criterion, a backsubstitution is performed to unfold T_ℓ, T_g, α, and the velocities for each phase. A call is then made to update all of the thermodynamic properties and their derivatives in preparation for the next time step.

Two options are available for solving the vessel pressure matrix. The first is a direct inversion of the pressure solution matrix when the number of vessel cells is sufficiently small (≤ 80). The second option is a coarse-mesh rebalance method for larger problems where Gauss-Seidel iteration is required. During the iteration the pressure solution is scaled nonuniformly to reduce the overall iteration error. Such scaling can be represented by

$$P'^{(i)} = S^{(i)} P^{(i)} \quad ,$$

where $P^{(i)}$ is the pressure solution vector after i iterations and $S^{(i)}$ is its scaling matrix that is diagonal with scalar elements s_j. We define a coarse-mesh region as those vessel regions having the same scale factor. The scaled pressure solution vector is then

$$P' = s_1 P_1 + s_2 P_2 + \cdots \quad ,$$

where P_i is a vector of pressures belonging to coarse-mesh region i. Use of this equation in the vessel pressure equation,

$$A \cdot P = B \quad ,$$

and the requirement that the least-squares error in P' be a minimum, then yield an equation for the scale factors,

$$U \cdot S = V \quad ,$$

where $U_{ij} = (P_i, AP_j)$, $V_i = (P_i, B)$, and the notation (X,Y) means the inner product of the vectors X and Y. The matrix size of U is equal to the number of coarse-mesh regions and is normally small enough to be solved by direct inversion.

The choice of coarse-mesh regions is extremely important. We have selected a scheme that follows the flow path in the vessel so that coarse-mesh regions are coupled in the flow direction. We use the facts that the vessel matrix A is a seven-stripe matrix for a three-dimensional vessel and that coupling occurs only between nearest neighbors (there is no coupling if neighbors are separated by a wall). Based on these facts we select coarse-mesh regions as follows. All mesh cells on a level in the downcomer (an annular region of downflow concentric with the core) form a single coarse-mesh region and all other mesh cells on a level form another coarse-mesh region. Hence, the total number of coarse-mesh regions is equal to the number of downcomer levels plus the total number of vessel levels. Although this choice of coarse-mesh regions is not unique, we have found it to be very effective in reducing the number of vessel iterations (typically a factor of 10).

The stability-enhancing two-step (SETS) method[3,4] used for one-dimensional flow eliminates the material Courant stability limit from all one-dimensional components. As a result, this upper bound to time step applies only in vessel cells and at junctions to the vessel.

The stability-enhancing two-step method consists of a basic step (that is almost identical to the standard semi-implicit method used in the vessel) and a stabilizing step. For homogeneous flow, the order of these steps does not matter. However, for two-fluid flow with noticeable relative velocity it is necessary to do the stabilizing step for the equations of motion before the basic step. When this stabilizing step precedes the basic step, an initial explicit prediction of velocities gives strong coupling through the interfacial drag terms without requiring direct communication between the stabilizing equations for liquid and vapor motion. To provide improved conservation and to minimize CDC 7600 machine storage required by TRAC, the stabilizing steps for mass and energy equations are done as the final portion of the calculation.

These basic numerical techniques have proven to be very effective for both short- and long-term transients. The implicit coupling of the mass and energy equations accomplished by the initial linearization produces a very sturdy scheme even when the phase change rate is high. The stabilizing procedure used in the one-dimensional components permits much larger time steps for slow transients (where accuracy considerations can allow a large timestep) than would otherwise be possible.

TRAC is a reasonably modular code. The physical components to be modeled (pipes, steam generators, pumps, etc.) have corresponding FORTRAN modules to perform the calculation. These modules, which set up the special requirements for a given component, call lower-level generic routines to solve for heat transfer and hydrodynamics.

At least six passes are made through each component. On the first pass heat-transfer coefficients are evaluated and the matrix elements for the one-dimensional stabilizer motion equations are obtained. A second call to the one-dimensional hydro routines produces a back substitution for the stabilizer momentum terms. The next two or more passes call the semi-implicit portion of the fluids subroutines until a solution within a convergence tolerance is obtained. A pass on the one-dimensional scalar stabilizer equations is performed followed by the final stabilizer backsubstitution, calculation of the wall and rod heat-transfer conduction solution, and the generation of all of the final quantities required for the next time step.

Timing Information

Typical computing time on a CDC 7600 for this cycle is 2.0 - 3.5 ms/cycle-fluid mesh cell. The times can vary depending on the ratio of three- to one-dimensional mesh cells, number of iterations (which is typically two to four, and the number of wall conduction nodes. Some preliminary studies have suggested that by redoing the data structure, switching to faster (but less accurate) system calls for exponentiation, trigonometric functions, etc., and performing some general code cleanup, the cycle time could be reduced 40-50%. Some of this work is going on but at a rather modest level.

On a fairly typical problem with one-dimensional loops and a three-dimensional coarsely noded vessel, the first two passes (termed a prepass) take about 23% of the CPU time, the semi-implicit hydro about 48%, the final two passes (post-pass) 16%. The heat transfer is performed during the pre- and postpasses and represents a large fraction of those compute times. Input and general overhead constitute the remainder. The bane of some hydro codes, the property routines, require only about 10-15% of the total time because TRAC uses rather fast polynomial fits.

The CRAY version runs approximately 2.3 times faster than the 7600 code. This represents only the speedup available for scalar operations. No serious effort has been yet made to take advantage of the vector abilities of the CRAY. The heart of the computation consists of approximately 10000 FORTRAN lines with many imbedded "if" tests. The basic numerical scheme is implicit, which further complicates the problem of producing efficient vector coding. While about 20% of the 10000 lines can be converted in a straightforward manner, this might at best produce a speedup of 20% (if the vector operations were infinitely fast). To vectorize the remainder of the code would require a great deal of effort that, to this point, has not been available.

Parallel and Distributed Processing

The fast scientific computers of the future will almost certainly be parallel devices in that several or many tasks can go on concurrently under the overall control of a host or central CPU. The amount of effort to convert TRAC to run efficiently on a multiple CPU machine appears to be much less than to vectorize the code, assuming of course that the higher-level software tools necessary for the job are available.

The structure of TRAC is already highly parallel. The most obvious examples of this are fluid and metal properties and the one-dimensional wall heat transfer. These items may be computed independently for all mesh cells in the system. The solution of the finite-difference fluid flow equations is also highly parallel despite its implicitness. The vast majority of the work for these equations is a cell-by-cell evaluation of coefficients for linearized equations. Parallelism in the TRAC hydrodynamics solution procedure exists at two basic levels. First, in a higher-level sense, operators may take advantage of the existing component structure in TRAC and calculate the individual components in parallel although some internal regrouping of components might be necessary to equalize the workload. This approach can utilize (very efficiently, as we show below) multiprocessor architectures that now, or will soon, consist of only a few processors. At a deeper level, all the mesh cells of a system may be computed in parallel. This approach requires many more processors and therefore is a longer-term prospect. In this section we first present our results for the higher (component) level scheme, then our results for the mesh cell scheme.

A. High-Level TRAC Parallelism

For running each TRAC component through individual processors in parallel, the main global (non-parallel) task is the solution of an N x N linear system for hydrodynamic quantities at the component junctions (such as pressure variations), where N is the number of junctions. This is achieved presently in TRAC by Gaussian elimination. Typically, this must be done 6-8 times per time step, depending on the number of iterations required.

With parallel and sequential coding identified, some quantitative estimates of the parallel performance can be made based on our CDC 7600 timings. Assume that a problem with n finite-difference mesh cells is to be computed. Let O_p be the floating point operation count per cell per cycle that can be performed in parallel mode and O_s be the count for operations to be done serially. The total operation count for the code is

$$O_T = O_p n + O_s \quad .$$

If S_{pp} is the inherent speed (in floating point operations per second) for a parallel processor unit, then the time required for a machine with N such units (each computing one or more TRAC components) to complete one cycle of calculation is

$$\tau_{pp} = \frac{O_p}{S_{pp}} (n/N) + \frac{O_s}{S_{pp}} \quad ,$$

where the ratio n/N is rounded up to the nearest integer. For a CDC 7600 the time would be

$$\tau_{76} = \frac{O_p n + O_s}{S_{76}} \quad ,$$

where S_{76} is the 7600 execution speed. The ratio R of the two times is

$$R = \frac{O_p(n/N) + O_s}{O_p n + O_s} \left(\frac{S_{76}}{S_{pp}}\right) .$$

This demonstrates the relation between the number of processors and processor speed required to achieve break even with a 7600. Another useful ratio is that of the time spent in serial mode to that in parallel computing

$$r_{sp} = \frac{O_s}{O_p(n/N)} .$$

For TRAC, with its current network junction matrix solution procedures, we have found this ratio to be

$$r_{sp} \approx \frac{0.001}{(n/N)} N^3 .$$

The N^3 dependence comes from the use of LU decomposition algorithms in the network junction matrix solution procedure. Other matrix solution procedures could improve this ratio for large N, but would require substantial recoding in TRAC. For current multiprocessors N is a relatively small number (2-8). As an example, on a four-processor machine computing a problem with a large number of cells, the ratio is

$$r_{sp} \approx \frac{0.3}{n} .$$

For this configuration the performance ratio to a 7600 becomes

$$R \approx \frac{1 + 0.3/n}{4 + 0.3/n} \left(\frac{S_{76}}{S_{pp}}\right) \quad .$$

Thus, a high degree of parallel efficiency is predicted.

B. Mesh Cell Parallel Simulation

As mentioned above, TRAC has potentially a deeper level of parallelism than that of the reactor components. The code spends much of its time looping over the individual computational mesh cells that constitute a component. For a given component, most of the hydrodynamics calculation is done in this mode; inspection of the code shows only a relatively small amount of FORTRAN coding devoted to nonparallel tasks, notably solution of tridiagonal equation systems that account for the coupling of each cell to its nearest neighbor.

We have obtained quantitative timing results, using a relatively small (~3500 lines) subset of the full TRAC code (the "TRAC benchmark code") to test the feasibility and efficiency of running individual mesh cells in parallel. The TRAC benchmark code is well suited to this task, as the code only loops over a small number of cells with no network junctions involved. Only TRAC's "basic" (semi-implicit) equation set is solved. The stabilizer equations of motion, mass, and energy, and all heat transfer, are not included. However, the stabilizer hydrodynamics step has a structure similar to the basic step. Therefore, the TRAC benchmark code provides a good test of the parallel efficiency of TRAC hydrodynamics computation at the mesh cell level. Heat transfer computations can consume a significant amount of TRAC time, depending on the problem being run. However, inspection of the coding shows the potential of achieving a high degree of heat-transfer parallel efficiency at the mesh cell level.

Our final results obtained with the TRAC benchmark on the Cray-1 showed that serial computation in the basic hydrodynamics, specifically the solution of a tridiagonal system for the pressure variations in each cell, accounted for 1% of total CPU time. The solution calculation time for the tridiagonal system is essentially proportional to the number of mesh cells. The 1% ratio should be invariant with respect to number of mesh cells in the component. Therefore, for the case of one processor dedicated to each cell, the ratio of serial to parallel computing time within a component is

$$r_{sp} = \frac{0.01n}{0.99} \approx .01n \quad ,$$

where n is the number of mesh cells. This indicates a good degree of parallel efficiency for TRAC hydrodynamics at the mesh cell level. For the specific results obtained here, the formula for parallel speedup calculation reduces to

$$\frac{1}{R} = \frac{1.0}{0.01 + 0.99/n} \quad ,$$

where we assume one processor per mesh cell and a basic CPU performance ratio of unity. It can be seen that the speedup has a limit of 100. The approach to this limit is shown by the following table:

n	speedup = 1/R
1	1.00
2	1.98
10	9.17
100	50.25
1000	90.99
1000000	99.99

It is apparent that parallel efficiency is expected to be excellent at 10 mesh cells and to fall below 50% only beyond 100 cells. It is notable that a typical TRAC one-dimensional component contains about 10 cells. As indicated above, we have not quantitatively studied the parallel efficiency of TRAC wall heat transfer and material property calculation, but it is expected to be similarly high.

A typical full-scale TRAC problem uses from 100-1000 mesh cells. We have determined here that abandoning the present TRAC networking scheme and going to an approach involving computation of the entire system in parallel, on an individual mesh cell basis, is most likely not an efficient approach (even not considering the considerable reprogramming effort that would be required). The large matrix that must be solved in serial mode will ruin parallel efficiency. However, it is a very attractive idea to run the mesh cells of individual TRAC one-dimensional components in parallel (~10 cells/component) as the components themselves are run in parallel (running parallel components is the original TRAC parallel concept discussed above). This would in effect split the large linear system into smaller parallel tridiagonal systems for each component. This approach would maximize parallel efficiency while preserving the current TRAC code structure.

The basic hydro routines are being benchmarked on a Dynelcor Heterogeneous Element Processor (HEP). The current machine has only a single CPU but up to eight tasks can be pipelined on the processor, giving the appearance of parallel operation. Future versions of the HEP will have multiple CPUs. The Los Alamos Computing Division is constructing a device with 16 CPUs and a user-controlled architecture that also will be used in TRAC benchmark studies.

C. Distributed Processing

A more immediate application of parallelism involves the TRAC Nuclear Plant Analyzer. This is a joint effort with The Idaho National Engineering Laboratory (INEL) and involves producing an interactive user interface for both TRAC and

RELAP5 (an INEL-developed thermal-hydraulic code). This interface will be a very flexible menu-driven system in which most of the I/O, including color graphics, will be performed on an intelligent workstation. The hydrodynamic and heat-transfer calculations still will be performed on a mainframe; however, by downloading the input processing and graphics, effective speed gains should be realized.

Although the project is still in its infancy, a trial version of the system should be available by mid-FY84. As both memory and CPU prices fall and array processors proliferate, it should be possible in a few years to consider running the entire calculation on a personal workstation.

5. SUMMARY AND CONCLUSIONS

TRAC is a representative large thermal-hydraulic code that performs calculations of multiphase internal flows. The implicit numerics allow large time steps but also make vectorization difficult. The basic numerical techniques are highly adaptable to the parallel architectures that will become more prevalent in the future. In addition, current effort is being directed toward downloading input processing and output graphics onto the current generation of Motorola 68000 based intelligent workstation.

REFERENCES

1. "TRAC-PF1, An Advanced Best-Estimate Computer Program for Pressurized Water Reactor Analysis," Los Alamos National Laboratory, to be published.
2. Liles, D. R., and Reed, W. H., "A Semi-Implicit Method for Two-Phase Fluid Dynamics," J. Of Comp. Physics 26, No. 3, 390-407 (1978).
3. Mahaffy, J. H., "A Stability-Enhancing Two-Step Method for One-Dimensional Two-Phase Flow," Los Alamos Scientific Laboratory report LA-7951-MS (1979).

4. Mahaffy, J. H., "A Stability-Enhancing Two-Step Method for Fluid Flow Calculations," submitted to J. Comp. Phys. [Los Alamos National Laboratory report LA-UR-81-1398 (October 1981)].

Supported by the Nuclear Regulatory Commission.

Safety Code Development Group
MS K553
Los Alamos National Laboratory
Los Alamos, NM 87545

ACCELERATION OF CONVERGENCE OF NAVIER–STOKES CALCULATIONS

R. W. MacCormack

1. Introduction

Currently unsteady Navier-Stokes calculations in three dimensions for high Reynolds number flows past relatively simple body geometries require thousands of time steps for convergence. If we are to use Navier-Stokes calculations in the practical design of complex aerodynamic shapes, as found for example in modern aircraft configurations, we must develop procedures to accelerate convergence. The present paper presents one such procedure based upon the multigrid theory devised by Brandt [1], extended by Ni [2] and Johnson [3] for the Euler and Navier-Stokes equations, and formulated by Denton [4] for explicit finite-volume calculations. The multigrid method to be presented herein is designed to accelerate the convergence of implicit methods applied to solve the Navier-Stokes equations at high Reynolds number.

ISBN 0-12-546080-5

2. The Navier-Stokes Equations

The unsteady compressible form of the Navier-Stokes equations in two-dimensions can be written in the following form

$$\frac{\partial U}{\partial t} = -\frac{\partial F}{\partial x} - \frac{\partial G}{\partial y} \tag{2.1}$$

where

$$U = \begin{pmatrix} \rho \\ \rho u \\ \rho v \\ e \end{pmatrix} \qquad F = \begin{pmatrix} \rho u \\ \rho u^2 + \sigma_x \\ \rho uv+\tau_{xy} \\ (e+\sigma_x)u+\tau_{xy}v - k\,\frac{\partial T}{\partial x} \end{pmatrix}$$

$$G = \begin{pmatrix} \rho v \\ \rho uv+\tau_{yx} \\ \rho v^2+\sigma_y \\ (e+\sigma_y)v + \tau_{yx}u - k\,\frac{\partial T}{\partial y} \end{pmatrix}$$

$$\sigma_x = p - \lambda\left(\frac{\partial u}{\partial x} + \frac{\partial v}{\partial y}\right) - 2\mu\,\frac{\partial u}{\partial x}$$

$$\tau_{xy} = \tau_{yx} = -\mu\left(\frac{\partial u}{\partial y} + \frac{\partial v}{\partial x}\right)$$

and

$$\sigma_y = p - \lambda\left(\frac{\partial u}{\partial x} + \frac{\partial v}{\partial y}\right) - 2\mu\,\frac{\partial v}{\partial y}$$

The above equations are written in terms of density ρ, x and y velocity components u and v, viscosity coefficients λ and μ, total energy per unit volume e, coefficient of heat conductivity k, and temperature T. Finally, the pressure p is related to the specific internal energy ε and ρ by an equation of state, $p(\varepsilon,\rho)$, where $\varepsilon = e/\rho - (u^2+v^2)/2$.

The Navier-Stokes equations can be written also in integral form. We first define

$$\vec{F} = F\vec{i}_x + G\vec{i}_y$$

and

$$\vec{\nabla} = \frac{\partial}{\partial x}\,\vec{i}_x + \frac{\partial}{\partial y}\,\vec{i}_y$$

where $\vec{i}_x$ and $\vec{i}_y$ are unit vectors in x and y coordinate directions. Equation (2.1) can then be written

$$\frac{\partial U}{\partial t} = -\,\vec{\nabla}\cdot\vec{F}$$

Integrating this equation over volume V we have

$$\int_V \frac{\partial U}{\partial t}\,dv = -\int_V \vec{\nabla}\cdot\vec{F}\,dv$$

and using the divergence theorem to replace the volume integration on the right with a surface integration we obtain

$$\int_V \frac{\partial U}{\partial t}\,dv = -\int_S \vec{F}\cdot\vec{n}\,ds \tag{2.2}$$

where the surface S encloses volume V.

3. Implicit Solution of the Navier-Stokes Equations

Equation (2.1) can be solved [5,6,and 7] by first discretizing the flow field into a two-dimensional set of mesh points separated in the x and y directions by Δx and Δy and then by following the three steps given below.

(i) $$\Delta U_{i,j} = -\,\Delta t\left(\frac{\Delta F}{\Delta x} + \frac{\Delta G}{\Delta y}\right)^n_{i,j} \tag{3.1}$$

(ii) $$\{I + \cdots\}\,\delta U_{i,j} = \Delta U_{i,j} \tag{3.2}$$

$$\text{(iii)} \quad U^{n+1}_{i,j} = U^{n}_{i,j} + \delta U_{i,j} \tag{3.3}$$

For the complete details see references [5,6, and 7]. Step (i) above is a local explicit approximation of the governing Equation (2.1) at each mesh point (i.j) at time $t = n\Delta t$. The predicted change in the solution given by $\Delta U_{i,j}$ is a fair approximation to the actual solution change if Δt is small enough. If Δt is large, as is desired to march the solution along in time efficiently, additional global information is required to approximate accurately the actual solution change. Step (ii) is an implicit equation for providing the additional information about global solution change needed at each mesh point to determine the local solution change for large time steps. The bracketed factor appearing in Equation (3.2) must be inverted before the implicitly predicted solution change given by $\delta U_{i,j}$ can be determined. Once determined, the new solution at $t = (n+1)\Delta t$ is obtained in step (iii).

The integral form of the governing equations, Equation (2.2), can be similarly solved by first discretizing the flow field into an arbitrary set of finite volumes (quadrilaterals in two-dimensional space) as shown in Figure 1. For each finite volume of the constructed mesh the average solution value is given by

$$\bar{U}_{i,j} = \frac{1}{V_{i,j}} \int_{V_{i,j}} U dv$$

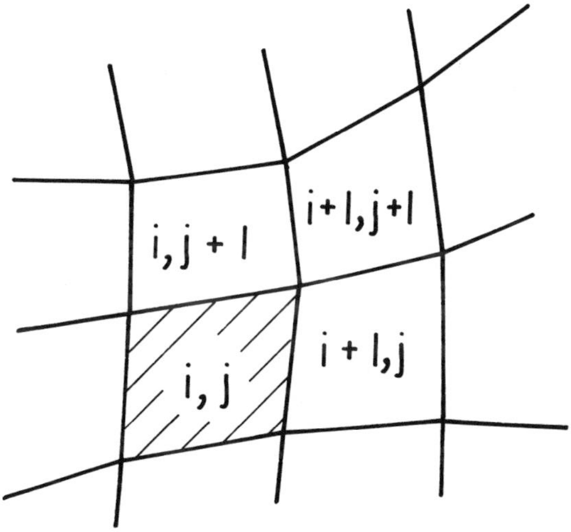

Figure 1. Sketch of finite-volume grid

where $V_{i,j}$ is the volume (area in two-dimensions) of finite volume (i,j). For each finite volume Equation (2.2) can be written as

$$\frac{\partial \bar{U}}{\partial t} = -\frac{1}{V}\int_S \vec{F}\cdot\vec{n}\,ds \tag{3.4}$$

and is solved implicitly by the following three steps

(i) $$\overline{\Delta U}_{i,j} = -\frac{\Delta t}{V_{i,j}}\{(\vec{F}\cdot S)_{i+1/2,j} + (\vec{F}\cdot\vec{S})_{i-1/2,j} + (\vec{F}\cdot\vec{S})_{i,j+1/2} + (\vec{F}\cdot\vec{S})_{i,j-1/2}\} \tag{3.5}$$

(ii) $$\{I + \cdots\}\,\delta U_{i,j} = \overline{\Delta U}_{i,j} \tag{3.6}$$

(iii) $$\bar{U}^{n+1}_{i,j} = \bar{U}^{n}_{i,j} + \delta U_{i,j} \tag{3.7}$$

The four terms of form $\Delta t(\vec{F}\cdot\vec{S})$ of Equation (3.5) represent fluxes of mass, momentum, and energy across the right, left, top, and bottom sides respectively, of finite volume (i,j). Equation (3.5) is written in strong conservation form: The flux leaving finite volume (i,j) across side $\vec{S}_{i+/2,j}$ is received exactly by finite-volume (i+i,j) and the summation $\sum_{i,j} \overline{\Delta U}_{i,j} V_{i,j}$ is composed of flux terms at the outer boundaries of the flow field only -- all others cancel internally. If the implicit Equation (3.6) is also written in strong conservation form then the change in the total mass, momentum, and energy within the calculated flow field, $\sum_{i,j} \delta U_{i,j}\, V_{i,j}$, is determined only by the flux flowing into or out of the flow volume across the outer mesh boundaries.

4. Convergence Considerations

An initial condition is required to start the calculation off at $t = 0$. For simplicity the sudden immersion of a body in a uniform flow field is often used. During the time march of the solution to convergence for a steady flow or through the time history of an unsteady one, the disturbances initially created at body surfaces propagate out away from the body to warn the surrounding flow to adjust for the bodies presence. Convergence for steady state flows or time accuracy independent of the choice of initial conditions for unsteady flow will occur only after the disturbances signifying the body's presence propagate throughout the region contained within the body's domain of influence. Figure 2 shows a mesh about a typical aerodynamic body in a compressible viscous flow. The mesh is highly stretched to resolve both the viscous effects occuring within a thin layer of thickness δ near the body and the predominantly inviscid flow within the domain of influence of the body that extends many body lengths away. The mesh spacing varies from $\Delta\eta_{min} << \delta$ deep within the thin boundary layer to $\Delta\eta_{max}$ of the order of the body length itself in the far field.

The choice of a time step size is more difficult. Before the development of noniterative implicit methods, time step sizes were severely limited according to stability criteria by the magnitudes of the characteristic velocities

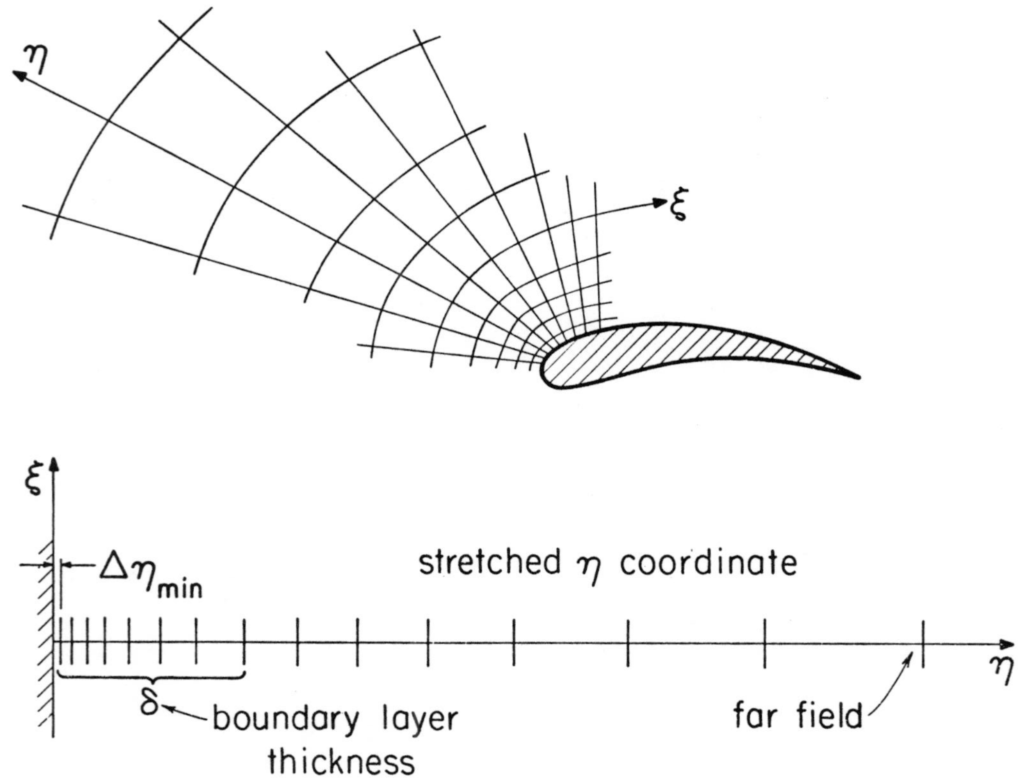

Figure 2. Mesh for a compressible viscous flow about an aerodynamic body.

and kinematic viscosities of the flow and the mesh point spacings occuring within the calculation. Time steps were forced to be chosen small enough to resolve all time scales possible within the flow field, for example,

$$\Delta t_{explicit} \leq \frac{\Delta\eta_{min}}{c} \tag{4.1}$$

where c represents the maximum characteristic signal speed of the flow. For information traveling at this slow pace, $\Delta\eta_{min}$ per time step, to reach distances several body lengths away would require an impractical large number of steps and amount of computer time.

The development of implicit methods removed the strict stability conditions on the choice of time step size and improved numerical efficiency in some cases by several orders of magnitude. For viscous flow calculations time step sizes are typically chosen of the order of

$$\Delta t_{implicit} \leq \frac{\delta}{c}\ , \tag{4.2}$$

where δ represents a small characteristic flow length, perhaps of the order of the boundary layer thickness, that is usually many times larger than $\Delta\eta_{min}$. This choice permits good temporal resolution near the body, but unfortunately still can represent a small pace for information traveling in the highly stretched mesh regions of the far field. Many time steps still would be required for information to travel even between nearest neighbors in the far field.

A procedure called "local time stepping" is currently receiving renewed interest for accelerating numerical convergence. According to this procedure the time step varies throughout the flow field and is chosen locally by

$$\Delta t_{\text{local time stepping}} \leqslant \frac{\Delta\eta}{c}\,, \tag{4.3}$$

where $\Delta\eta$ is the local mesh point spacing. The main advantage of this procedure is that information can travel from each mesh point to its nearest neighbors during each time step of the calculation. The main disadvantage is that the solution evolves on a warped time surface, which can cause severe difficulties in analyzing unsteady phenomena. However, for steady state calculations these difficulties are of little importance.

The multigrid theory originally devised by Brandt [1], extended by Ni [2] and Johnson [3] for the Euler and Navier-Stokes equations, formulated by Denton [4] for explicit finite-volume calculations, and presented herein for use with implicit methods offers potentially greater numerical efficiencies than the procedures just discussed. During each step of the calculation the solution can be advanced simultaneously on several different grids, from fine to coarse, with a different time step chosen for each grid. Time steps can be chosen on the coarser grids large enough so that during each step information can travel distances far beyond the distances between nearest neighbors of the finer grids everywhere throughout the flow field. It is possible for information to travel between the far field and the boundary layer regions of the flow, like that sketched in Figure 2, within a single step of the calculation. We will illustrate the multigrid procedure for accelerating the convergence of first an explicit numerical method and then of an implicit one.

5. Multigrid-Explicit Method

First a mesh, like that shown in Figure 2, is chosen to resolve all the significant features of the flow and is called the fine mesh. Then the solution changes at each mesh point or finite volume are calculated explicitly using equations of the form of Equation (3.1) or Equation (3.5). The complete evaluation of the flux terms, F and G of Equation (3.1) or $(\vec{F}\cdot\vec{S})$ of Equation (3.5), including both their inviscid and viscous parts are determined at each mesh location. The time step for this calculation is chosen small enough so that the strict explicit stability condition is satisfied everywhere throughout the mesh. Equation (3.2) and Equation (3.6) is then replaced by

$$\delta U^{(1)}_{i,j} = \Delta U_{i,j} \tag{5.1}$$

or

$$\delta U^{(1)}_{i,j} = \overline{\Delta U}_{i,j} \, . \tag{5.2}$$

respectively. The superscript (1) indicates that the above solution changes are determined for points of the first or fine mesh.

A coarse mesh is then formed by deleting a set of mesh surfaces from the original fine mesh, as shown in Figure 3. For example, if every other surface is removed, a new coarse two-dimensional mesh would contain only one fourth (one eighth in three dimensions) as many points or finite volumes as before, each containing four mesh volumes of the previous mesh. Let the indices $i^{(2)}$ and $j^{(2)}$ represent the indices of the new coarse grid. We define the solution change, $\delta U^{(2)}_{i^{(2)},j^{(2)}}$, on the new coarse grid obtained by deleting every other surface of the previous grid, as follows

$$\delta U^{(2)}_{i^{(2)},j^{(2)}} = 2\{V^{(1)}_{i,j}\delta U^{(1)}_{i,j} + V^{(1)}_{i+1,j}\delta U^{(1)}_{i+1,j} + V^{(1)}_{i,j+1}\delta U^{(1)}_{i,j+1}$$

$$+ V^{(1)}_{i+1,j+1}\ \delta U^{(1)}_{i+1,j+1}\}/\{V^{(1)}_{i,j} + V^{(1)}_{i+1,j} + V^{(1)}_{i,j+1} + V^{(1)}_{i+1,j+1}\},$$

where the four mesh cells of the original fine grid (i,j), (i+1,j), (i,j+1) and (i+1,j+1) are contained in new coarse mesh cell $(i^{(2)}, j^{(2)})$ and $V^{(1)}_{i,j}$ equals either $\Delta x \Delta y$ for the finite difference approach, Equations (3.1)-(3.3) or as defined previously for the finite-volume approach, Equations (3.5)-(3.7). The factor 2 appearing on the right hand side of the above equation indicates that the time step size for this solution change is twice that for the fine mesh, $2\Delta t^{(1)}$. Note that no flux terms, either viscous or inviscid, needed to be calculated to obtain the new coarse

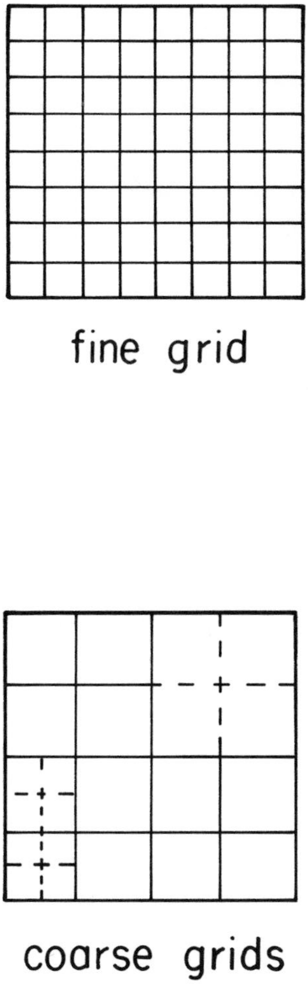

Figure 3. Coarse grid formation

mesh solution changes. Only a simple volume weighted averaging was required. Yet because of the strong conservation form of Equation (3.1) and Equation (3.5) the weighted sum exactly represents the net flux entering and leaving coarse mesh cell $(i^{(2)},j^{(2)})$ in terms of fluxes calculated on the original fine grid. For example, for the finite volume approach

$$\delta U^{(2)}_{i^{(2)},j^{(2)}} = -2\Delta t \{(\vec{F}\cdot\vec{S})_{i+3/2,j} + (\vec{F}\cdot\vec{S})_{i+3/2,j+1} +$$

$$(\vec{F}\cdot\vec{S})_{i-1/2,j} + (\vec{F}\cdot\vec{S})_{i-1/2,j+1} +$$

$$(\vec{F}\cdot\vec{S})_{i,j+3/2} + (\vec{F}\cdot\vec{S})_{i+1,j+3/2} +$$

$$(\vec{F}\cdot\vec{S})_{i,j-1/2} + (\vec{F}\cdot\vec{S})_{i+1,j-1/2}\}/$$

$$\{V^{(1)}_{i,j} + V^{(1)}_{i+1,j} + V^{(1)}_{i,j+1} + V^{(1)}_{i+1,j+1}\}$$

The direct advantage of the weighted averaging is that the fine scale flux resolution is retained on the succeeding coarse grid calculations without ever having to recalculate a flux term.

The generation of coarser grids from finer ones can be continued until perhaps a grid is obtained containing only one mesh cell. We define the solution change of each new coarse grid cell $(i^{(k)}, j^{(k)})$ in terms of the solution changes of the previous finer grid cells contained within as follows, for $k > 1$.

$$\delta U^{(k)}_{i^{(k)},j^{(k)}} = \frac{T^{(k)}}{V^{(k)}_{i^{(k)},j^{(k)}}} \sum V^{(k-1)}_{i^{(k-1)},j^{(k-1)}} \delta U^{(k-1)}_{i^{(k-1)},j^{(k-1)}} \qquad (5.3)$$

where

$$V^{(k)}_{i^{(k)},j^{(k)}} = \sum V^{(k-1)}_{i^{(k-1)},j^{(k-1)}}$$

and

$$\tau^{(k)} \leq \left[\frac{\min \{V^{(k)}_{i^{(k)},j^{(k)}}\}}{\min \{V^{(k-1)}_{i^{(k-1)},j^{(k-1)}}\}} \right]^{1/d}$$

The summations above contain all terms with indices $i^{(k-1)}$ and $j^{(k-1)}$ that represent mesh cells $(i^{(k-1)}, j^{(k-1)})$ geometrically contained within mesh cell $(i^{(k)}, j^{(k)})$. The factor $\tau^{(k)}$ represents the ratio $\Delta t^{(k)}/\Delta t^{(k-1)}$ and is determined as shown above with d equal to the number of spacial dimensions. Again, only a simple volume-weighted averaging is required to determine the solution changes of the next level grid and yet the flux resolution of the finest grid is retained.

If $K-1$ additional coarse grids are formed, the corresponding solution changes on each, determined by Equation (5.3), then need to be partitioned back to the original fine mesh cells. The new updated solution, in place of that given by Equation (3.3) or Equation (3.7), becomes

$$U^{n+1}_{i,j} = U^{n}_{i,j} + \delta U^{(1)}_{i,j} + \alpha \sum_{k=2}^{K} \delta U^{(k)}_{i^{(k)},j^{(k)}} \qquad (5.4)$$

The summation contains $K-1$ terms, each representing a solution change for a cell $(i^{(k)}, j^{(k)})$ containing the fine grid cell (i,j).

The factor α is chosen so that the physical domain of dependence is contained within the numerical domain even if the last grid contains only a single cell. The total time that the fine grid solution is advanced during each multigrid step is given by

$$\Delta T = \Delta t^{(1)} + \alpha(T^{(2)}\Delta t^{(1)} + T^{(3)}\,T^{(2)}\Delta t^{(1)}$$

$$+ \ldots T^{(k)}T^{(k-1)} \ldots T^{(2)}\Delta t^{(1)})$$

or

$$\Delta T = \{1 + \alpha \sum_{k=2}^{K} (\prod_{m=2}^{k} T^{(m)})\}\,\Delta t^{(1)} \tag{5.5}$$

Let's consider the following example to illustrate the application of the just described multigrid procedure. From an original fine equispaced grid containing 32 x 16 cells we can form the following four additional coarse grids.

k		GRID	TIME STEP
1	FINEST	32X16	Δt
2	COARSER GRIDS	16X8	$2\Delta t$
3		8X4	$4\Delta t$
4		4X2	$8\Delta t$
5		2X1	$16\Delta t$

The cell volumes increase by a factor of four on each new coarse grid obtained by deleting every other mesh line from the previous finer two dimensional grid. The factor $T^{(k)}$ equals two and the summation in Equation (5.3) contains four terms for each k. The last grid contains only two cells, each $16\Delta\eta$ wide where $\Delta\eta$ is the fine mesh point spacing. The centroids of these cells lie $8\Delta\eta$ away from the mesh boundaries and $8\frac{1}{2}\,\Delta\eta$ from the centroids of the boundary cells of the original fine mesh that lie just outside the interior mesh boundary. The largest total time that the flow volume can be advanced during each multigrid step, without the physical domain of dependence exceeding the numerical domain, is (see also Equation (4.1))

$$\Delta T_{max} \simeq \frac{8\frac{1}{2}\,\Delta\eta}{c} \simeq 8\tfrac{1}{2}\,\Delta t^{(1)}$$

From Equation (5.5)

$$\Delta T = \{1 + \alpha \sum_{k=2}^{5} 2^{k-1}\}\Delta t^{(1)}$$

$$= \{1 + 30\alpha\}\Delta t^{(1)}$$

Thus, for $\Delta T \leqslant \Delta T_{max}$, α must be less than or equal to $\frac{1}{4}$.

From Equations (5.3) and (5.4) we can determine the weight given to $\delta U^{(1)}_{i,j}$ in determining $U^{n+1}_{i,j}$. It is

$$1 + \alpha \sum_{k=2}^{K} \left(\prod_{m=2}^{k} \tau^{(m)}\right) \frac{V^{(1)}_{i,j}}{V^{(k)}_{i^{(k)},j^{(k)}}} \tag{5.6}$$

For our example above, the weight given to $\delta U^{(1)}_{i,j}$ is, with $\alpha = \frac{1}{4}$,

$$1 + \tfrac{1}{4} \sum_{k=2}^{K} 2^{(k-1)} \frac{1}{4^{(k-1)}}$$

$$= 1 + \tfrac{1}{4} \sum_{k=2}^{K} \left(\frac{1}{2}\right)^{k-1} \leqslant 1 + \tfrac{1}{4}$$

It appears possible with such a weight that the calculation could be unstable unless perhaps the CFL number (ratio of Δt used to Δt_{max}) is chosen less than 4/5 so that the product of the CFL number and the weight is less than one. This may be true. However, the wave length most often associated with instability is of length $2\Delta\eta$ on the fine mesh and this length is not supported on the coarser meshes. Also the weighted averaging procedure used to determine the new coarse mesh solution changes severely discriminates against this solution component.

6. Multigrid-Implicit Method

We will now try to combine the efficiency of implicit methods with the convergence acceleration properties of the multigrid procedure. Implicit procedures can be used to great advantage in computing viscous flows at high Reynolds number. The mesh must be fine near body surfaces to resolve the viscous layer and yet cover a large volume of the flow away from the body as well. Such a fine mesh is sketched in Figure 4a. To use the multigrid procedure we begin as before except we now use both Equations (3.1) and (3.2) or Equations (3.5) and (3.6) with the time step chosen as in Equation (4.2) to calculate $\delta U_{i,j}$ everywhere on the fine mesh. With this choice of time step $c\Delta t$ is of the order of δ instead of $\Delta\eta$ as before. To get the procedure started we will need to remove many more surfaces than just every other one in the boundary layer regions of the mesh to form volumes with linear dimensions of at least length δ. For the mesh sketched in Figure 4a with the stretched mesh direction indexed by j, let's assume that all the volumes except those within the boundary layer are already of linear dimensions of at least δ. We then define $\delta U^{(1)}_{i^{(1)},j^{(1)}}$ by

a)

b)

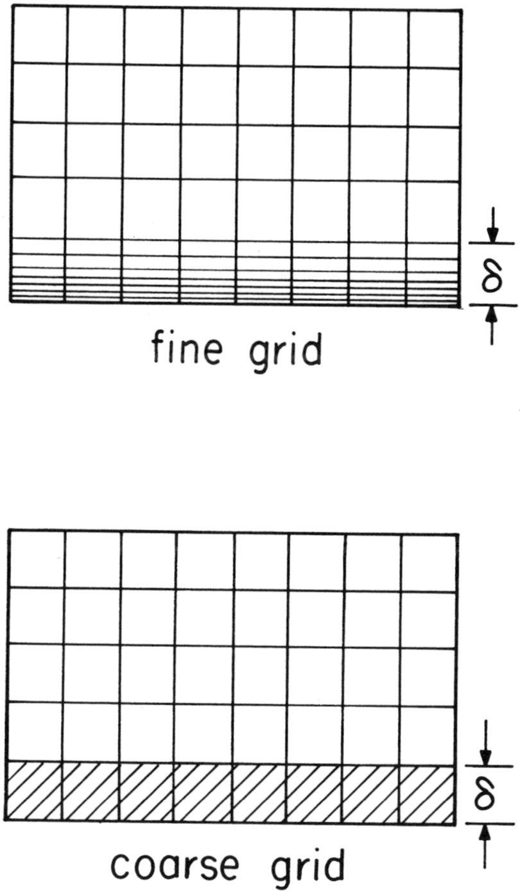

Figure 4a). Fine grid for an implicit calculation of a viscous flow and
b). a coarse grid formed to start multigrid procedure.

$$\left.\begin{aligned}
\delta U^{(1)}_{i^{(1)},j^{(1)}} &= \begin{cases} \delta U_{i,j}, & \text{if } j \text{ is outside the boundary layer} \\ \sum_{j\varepsilon\delta} V_{i,j}\delta U_{i,j}/V^{(1)}_{i^{(1)},j^{(1)}}, & \text{otherwise} \end{cases} \\
&\text{and} \\
V^{(1)}_{i^{(1)},j^{(1)}} &= \begin{cases} V_{i,j}, & \text{if } j \text{ is outside the boundary layer} \\ \sum_{j\varepsilon\delta} V_{i,j}, & \text{otherwise} \end{cases}
\end{aligned}\right\} \qquad (6.1)$$

thus we have obtained a set of solution changes and a new coarse mesh, Figure 4b, with spacings at least of length δ and a time step such that $c\Delta t^{(1)} \leqslant \delta$. This is essentially the condition we started with in the last section. We can now proceed as before using Equation (5.3) to determine solution changes on successively coarser grids. However, in place of Equation (5.4) we use the following to determine the new updated fine mesh solution.

$$U^{n+1}_{i,j} = U^{n}_{i,j} + \delta U_{i,j} + \alpha \sum_{k=2}^{K} \delta U^{(k)}_{i^{(k)},j^{(k)}} \qquad (6.2)$$

Note that the term $\delta U_{i,j}$ appears with no superscript indicating that the values determined by Equations (3.2) or (3.6) are used here instead of those obtained from Equation (6.1) to avoid loss of local resolution. The weight of this term in the above equation is

$$1 + \alpha \sum_{k=2}^{K} \left(\prod_{m=2}^{k} T^{(m)} \right) \frac{V_{i,j}}{V^{(k)}_{i^{(k)},j^{(k)}}} \tag{6.3}$$

with the volume $V_{i,j}$ without superscript indicating it is a volume of the original fine mesh. For volumes deep within the stretched boundary layer mesh the volume ratios in Equation (6.3) can be insignificantly small. This indicates some loss of resolution caused by the initial boundary layer averaging. Each $\delta U_{i,j}$ of the fine mesh is calculated stably with time step Δt by an implicit method, yet those in the boundary layer are disadvantageously weighted in Equation (6.1) in determining the first coarse mesh solution change. This loss of resolution can be restored as follows. Let

$$D^{(1)}_{i^{(1)},j} = \begin{cases} \delta U_{i,j} - \delta U^{(1)}_{i^{(1)},j^{(1)}} & \text{for } j \in \delta \\ 0 \quad , & \text{otherwise} \end{cases}$$

and

$$D^{(k)}_{i^{(k)},j} = \frac{D^{(k-1)}_{i^{(k-1)},j} \; V^{(k-1)}_{i^{(k-1)},j^{(k-1)}}}{V^{(k)}_{i^{(k)},j^{(k)}}}$$

for $k > 1$, $j \in \delta$, and with $V^{(k)}_{i^{(k)},j^{(k)}}$ as defined previously. The quantities defined above represent the difference between the solution changes in the boundary layer region of the fine mesh taken individually with full weights and the averaged values as used in the described multigrid procedure. Note that from Equation (6.1)

$$\sum_{j\varepsilon\delta} V_{i,j} D^{(1)}_{i^{(1)},j} = 0$$

The lost resolution is restored now if, instead of using Equation (6.2) to update the fine mesh solution, we use

$$U^{n+1}_{i,j} = U^{n}_{i,j} + \delta U_{i,j} + \alpha \sum_{k=2}^{k} (\delta U^{(k)}_{i^{(k)},j^{(k)}} + D^{k}_{i^{(k)},j^{(k)}}) \quad (6.4)$$

The weight of $\delta U_{i,j}$ in Equation (6.4) is now given by Equation (5.6) with $V^{(k)}_{i^{(k)},j^{(k)}}$ and $T^{(k)}$ given by Equations (6.1) and (5.3).

7. Results

The multigrid procedure just presented was applied to accelerate the convergence of an implicit method [7] for solving the unsteady Navier-Stokes equations. The methods were tested on a shock-wave boundary layer interaction problem. The flow field representing this interaction is sketched in Figure 5. The sketch shows an externally generated shock-wave incident upon a boundary layer on a flat plate. If the shock wave is strong enough the boundary layer will, as shown in the sketch, separate from the surface of the plate and reattach downstream. Between the separation and reattachment points there is a region of rotating fluid that causes the boundary layer to thicken and generate a series of compression and expansion waves that eventually form the reflected shock wave. The separation region is fairly sensitive to calculate and therefore serves as a good test for a numerical method.

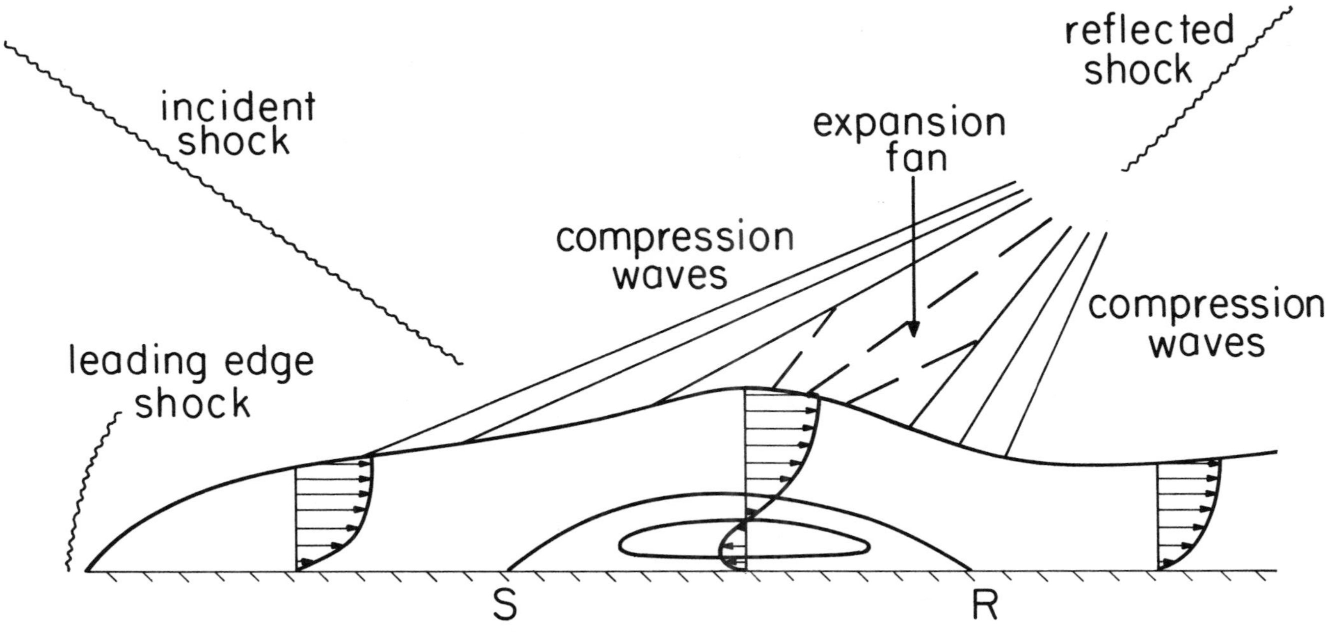

Figure 5. Sketch of shock-wave boundary-layer interaction

The initial flow field and boundary conditions are shown schematically in Figure 6. The initial condition for the interior of the flow field is uniform flow. The flow variables at the top mesh boundary were set to either free-stream values or values for a given shock strength so that the shock wave would impinge on the plate surface at the lower boundary at a given point. The lower surface was either a plane of symmetry or a wall surface and reflective boundary conditions were used. For the selected test case problem the free stream was supersonic and therefore the values at the upstream boundary were held fixed. At the downstream or exit boundary zero-order extrapolation ($U_{I,j} = U_{I-1,j}$) was used. The flow leaving is either parabolic (near the plate) or hyperbolic (away from the plate) in the streamwise direction so that errors made here should not propagate upstream. Tests using more accurate boundary treatment have verified that this is true.

The mesh contained 32x32 cells with 16 spanning the boundary layer. The mesh outside the boundary layer was uniform. Five additional grids were formed for the multigrid procedure just described containing 32x16, 16x8, 8x4, 4x2 and 2x1 cells. The time step chosen for the implicit method was such that $c\Delta t^{(1)} \leq \delta$, where δ was approximately the height of the fine boundary layer mesh. The factor $\tau^{(k)}$ was approximately two for $k = 2$ to 5 and $\alpha = 1/4$. The total time the solution was advanced each step by the multigrid procedure was approximately $8\tfrac{1}{2}\,\Delta t^{(1)}$.

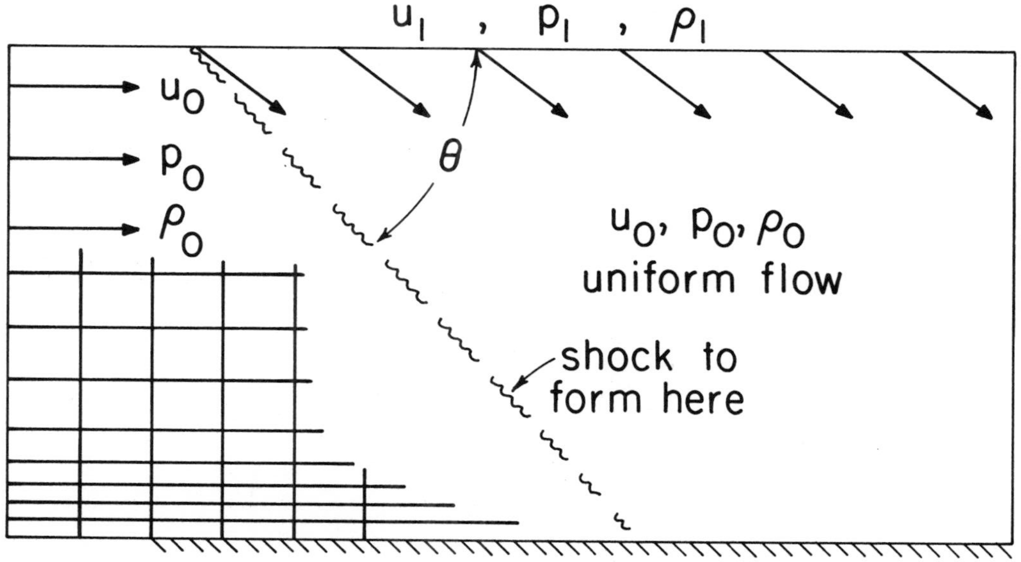

Figure 6. Initial flow field

The initial conditions of Mach two flow at a Reynolds number of 2.96×10^5 with a shock wave strong enough to increase surface pressure by a factor of 1.4 and cause boundary layer separation corresponds to an experimental case of Hakkinen [8]. The results of the implicit method compare well with experiment for this case [7] and are not repeated here. In Figure 7 the surface pressures at several time steps for the implicit method with the multigrid calculation are compared with that of the implicit method alone. After 32 time steps using only the implicit method the surface is still unaffected by the oncoming shock wave. While for the multigrid procedure information concerning the shock wave, though severely smeared, travels across the entire flow field in just a few steps and begins to show some correct form after 32 steps. The surface pressure at a fixed x location versus time step is shown in Figure 8 for both methods. The multigrid procedure appears to accelerate convergence in this figure by a factor greater than three. Figures 9 and 10 show the results of convergence at late times for both surface pressure and skin friction. Multigrid convergence acceleration appears here to be only approximately a factor of two.

The question arises of why the full convergence acceleration potential of $8\frac{1}{2}$ times was not realized for this case. The multigrid procedure did bring information concerning the shock at the top mesh boundary to the lower surface rapidly, perhaps more than eight times faster than the implicit method without multigrid. However the pacing item for this flow was the development of the boundary layer

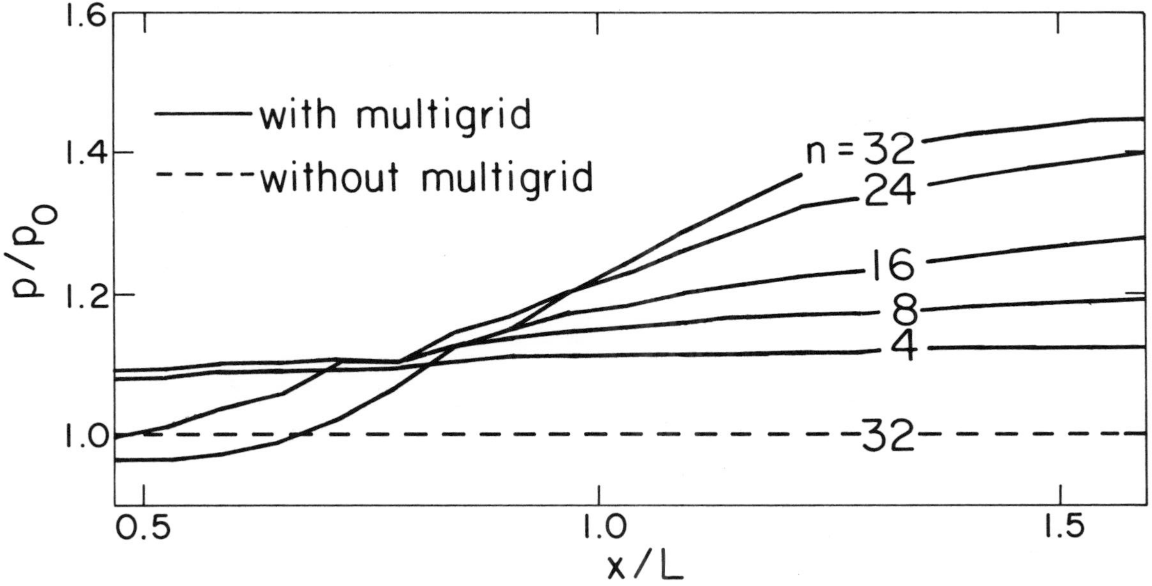

Figure 7. Surface pressure comparisons with and without multigrid at several time steps.

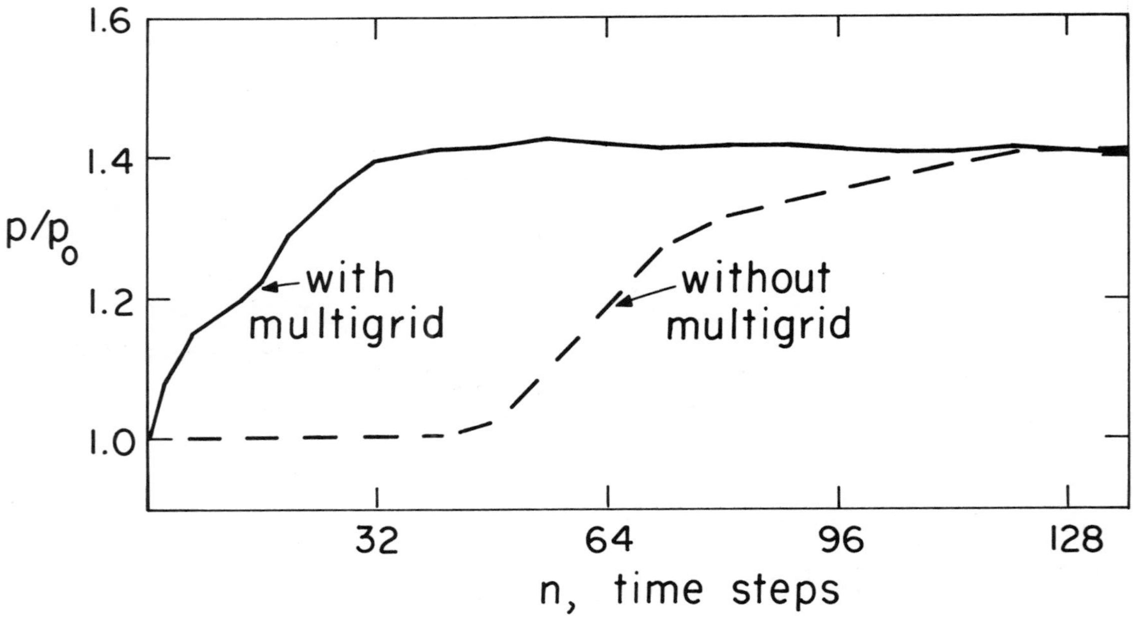

Figure 8. Surface pressure at a fixed x location versus time step with and without multigrid.

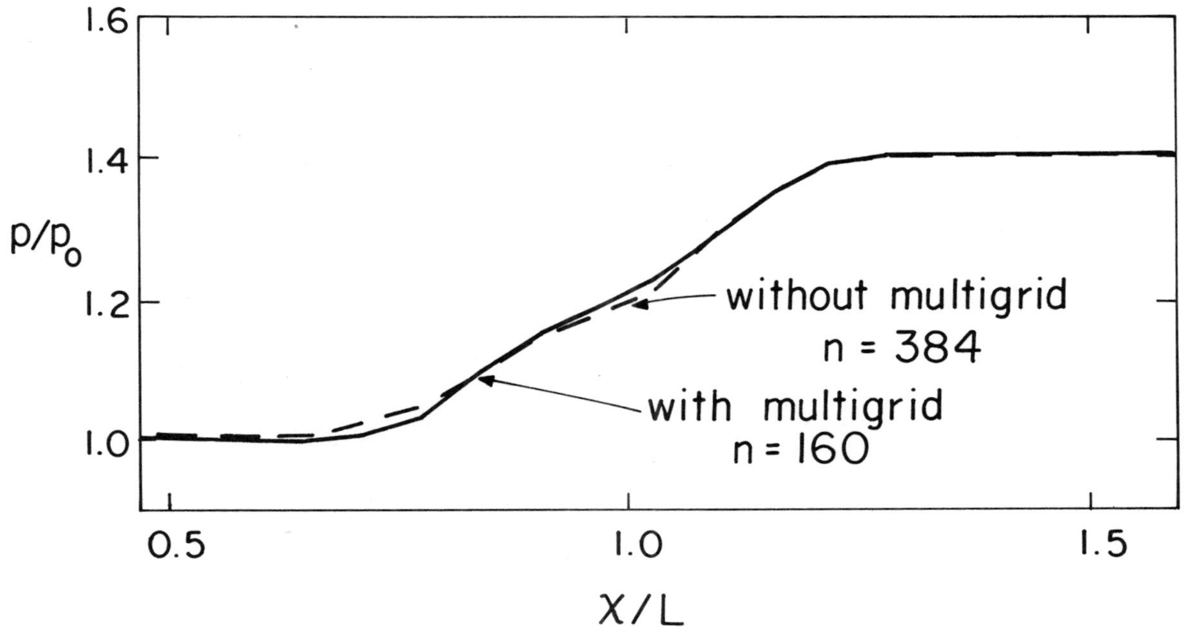

Figure 9. Surface pressure near convergence.

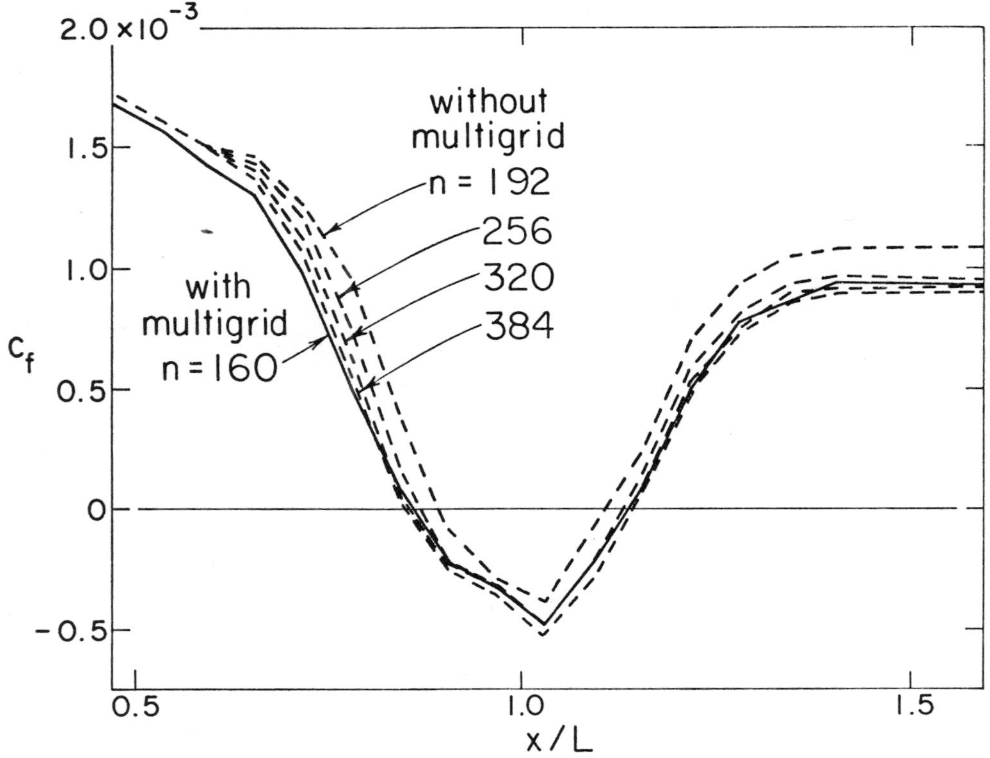

Figure 10. Skin friction near convergence.

itself from a condition of uniform flow. The additional information concerning the inviscid flow exterior to the boundary layer could do little to accelerate this development. The outer mesh was uniform and the top boundary was located relatively near the lower surface. For flows with outer boundaries far from body surfaces with highly stretched meshes like that shown in Figure 2 perhaps the situation would be reversed, the boundary layer would develop early and await the influence of the far field. For this latter case the multigrid procedure could perhaps achieve its full potential for convergence acceleration.

Nevertheless a respectable convergence acceleration of more then two was achieved. The additional cost per time step of adding the multigrid procedure to an existing implicit computational program was only approximately fifteen percent additional computer time. This was more than made up by the reduction in the number of time steps required for convergence. No additional memory was required for implementation. The procedure was called as a subroutine after the implicit solution changes were calculated and used storage arrays no longer needed by the existing program.

8. Conclusion

A multigrid method has been presented for accelerating the convergence of compressible unsteady Navier-Stokes calculations using implicit methods. The procedure does not depend on the particular form of the implicit method used. It is designed to retain the resolution of the finest grid, preserve strong conservation form, require no additional memory, and be added on to existing implicit programs for calculating compressible viscous flow at high Reynolds numbers.

References

1. Brandt, A. "Multigrid Solutions to Steady-State Compressible Navier-Stokes Equations I", Proceedings of the 5th International Symposium on Computing Methods in Applied Science and Engineering, Glowinski, R. and Lions, J. L., Editors, Versailles, France 1981.

2. Ni, R. H. "A Multiple Grid Scheme for Solving the Euler Equations", AIAA paper No. 31-1025, 1981.

3. Johnson, G. H. "Multiple Grid Acceleration of Lax-Wendroff Algorithms", NASA TM 82843, 1982.

4. Denton, J. D. "An Improved Time Marching Method for Turbomachinery Flow Calculation", ASME Paper No. 82-GT-239, 1982.

5. Briley, W. R, and McDonald, H. "Solution of the Three-Dimensional Compressible Navier-Stokes Equations by an Implicit Technique", Proceeding of the 4th International Conference on Numerical Methods in Fluid Dynamics, Springer 1975.

6. Beam, R. M. and Warming, R.F. "An Implicit Factored Scheme for the Compressible Navier-Stokes Equations", AIAA Paper No. 77-645, 1967.

7. Mac Cormack, R. W. "A Numerical Method for Solving the Equations of Compressible Viscous Flow", AIAA Paper No. 81-0110, January 1981.

8. Hakkinen, R. J., Greber, I., Trilling, L., and Abarbanel, S. S. "The Interaction of an Oblique Shock Wave with a Laminar Boundary Layer", Memo 2-18-59W, 959, NASA.

This work was supported by and performed at ICASE, NASA Langley Research Center, June 1982.

Department of Aeronautics and Astronautics, FS-10
University of Washington
Seattle, Washington 98195

ERROR ANALYSIS AND DIFFERENCE EQUATIONS ON CURVILINEAR COORDINATE SYSTEMS

C. Wayne Mastin

1. INTRODUCTION.

A computational grid must be constructed when solving partial differential equations by finite-difference or finite element methods. Presently there are many grid generation algorithms. The choice of algorithm will depend on the users desire for control over properties such as orthogonality of coordinate lines, location of grid points, and smoothness of grid point distribution. All of these properties may affect the accuracy of the numerical solution. A survey of grid generation techniques may be found in the article by Thompson et. al [7]. This report will deal with methods for deriving difference equations on curvilinear coordinate systems and the effect of coordinate systems on the solution. Recent contributions dealing with the effect of the grid on truncation error for one-dimensional problems have been made by Hoffman [3] and Vinokur [8].

The motivation for this investigation can be seen by considering some current problems in computational fluid dynamics. Available computers can be used to model the flow about a wing-fuselage configuration. When additional components, such as fins, stores, and nacelles are added to the aircraft, the computational region becomes increasingly complicated. The grid can be extremely distorted and special difference formulas may be needed due to irregular neighborhood structures as encountered by by Lee et. al. [4] and Roberts [5]. In an attempt to limit grid distortion, overlapping grids have also been used for complicated regions. Each component is endowed with its own local coordinate system, and interpolation is used in the solution algorithm. Various interpolation procedures have been used by Atta [1], Atta and Vadyak [2], and Starius [6]. The

ISBN 0-12-546080-5

possible impact of the interpolation procedure on the solution algorithm will be investigated. It is noted that the interpolation technique may effect the local truncation error as well as the convergence rate of iterative algorithms and the stability of explicit algorithms.

2. FINITE-DIFFERENCE EQUATIONS.

A curvilinear coordinate system in the xy-plane is understood to be the image of a rectangular Cartesian coordinate system in a $\xi\eta$-plane. The induced grid is therefore composed of quadrilateral cells and difference equations may be derived by transforming the partial differential equation to the $\xi\eta$-plane.

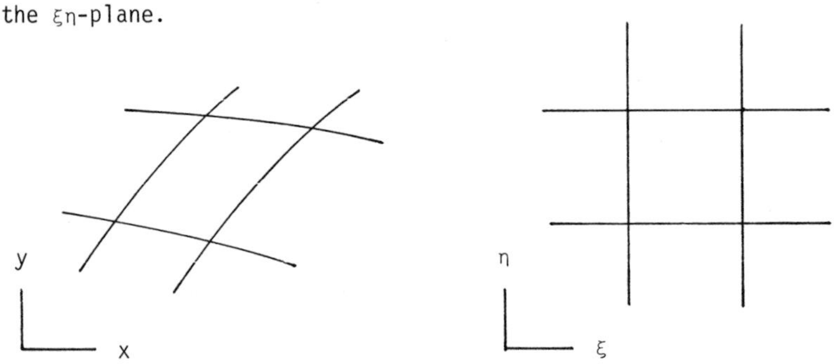

Figure 1.

A typical grid cell is indicated in Figure 1. This derivation gives no information on local truncation error, so an alternate derivation will be presented. Regardless of the derivation, the difference equations will involve derivatives of x and y with respect to ξ and η. Since the grid may be given only as a set at data points, it will be assumed that these derivatives are approximated using differences. It can be shown that the exact computation of these derivatives does not increase the accuracy of the method.

Second order central differences are commonly used in the numerical solution of partial differential equations. The truncation error depends on the grid spacing in the curvilinear coordinate system, and therefore, the corresponding grid spacing in the $\xi\eta$-plane will at present be assumed unity. The difference approximations with respect to ξ and η are then

$$\begin{aligned} f_\xi &= (f(\xi+1,\eta) - f(\xi-1,\eta))/2 \\ f_\eta &= (f(\xi,\eta+1) - f(\xi,\eta-1))/2 \\ f_{\xi\xi} &= f(\xi+1,\eta) + f(\xi-1,\eta) - 2f(\xi,\eta) \\ f_{\xi\eta} &= (f(\xi+1,\eta+1) + f(\xi-1,\eta-1) - f(\xi-1,\eta+1) - f(\xi+1,\eta-1))/4 \\ f_{\eta\eta} &= f(\xi,\eta+1) + f(\xi,\eta-1) - 2f(\xi,\eta). \end{aligned} \tag{1}$$

The local truncation error in using these differences to approximate the derivatives with respect to x and y is revealed by examining a series expansion of the above differences at $(x(\xi,\eta), y(\xi,\eta))$. First derivative approximations are much simpler, and they will be considered first. After a little algebra, the difference expression f_ξ can be represented as

$$f_\xi = x_\xi \frac{\partial f}{\partial x} + y_\xi \frac{\partial f}{\partial y} + \frac{1}{2} x_\xi x_{\xi\xi} \frac{\partial^2 f}{\partial x^2} + \frac{1}{2}(x_\xi y_{\xi\xi} + y_\xi x_{\xi\xi}) \frac{\partial^2 f}{\partial x \partial y}$$

$$+ \frac{1}{2} y_\xi y_{\xi\xi} \frac{\partial^2 f}{\partial y^2} + \text{HOT}. \qquad (2)$$

The terms in this expansion can be separated into three categories. The first order terms are used in deriving the difference equations. The second order terms are due to the nonuniform spacing and curvature of the curvilinear coordinate systems. The remaining higher order terms (HOT) are proportional to the third power of the grid spacing, and terms of this order would appear even if a uniform rectangular grid were used. Clearly the same remarks can be made about the series expansion for f_η. From these comments it follows that one condition for the derived difference approximations to be second order, in the sense that the local truncation error is the same order as the square of the grid spacing, is the condition that the following quotients

$$\frac{x_{\xi\xi}}{|r_\xi|^2}, \quad \frac{y_{\xi\xi}}{|r_\xi|^2}, \quad \frac{x_{\eta\eta}}{|r_\eta|^2}, \quad \frac{y_{\eta\eta}}{|r_\eta|^2}, \quad r = (x,y) \qquad (3)$$

remain bounded as $|r_\xi| + |r_\eta|$ approaches zero. A second condition arises when examining the form of the truncation error in the approximation of $\frac{\partial f}{\partial x}$ and $\frac{\partial f}{\partial y}$. From (2) it is seen that the truncation error for $\frac{\partial f}{\partial x}$ can be written

$$Tx = \frac{1}{J}(y_\eta T\xi - y_\xi T\eta)$$

where

$$J = x_\xi y_\eta - x_\eta y_\xi$$

and $T\xi$ and $T\eta$ are the second and higher order terms in (2) and the analogous expansion for f_η. Certainly some lower bound on the rate at which J approaches zero is necessary. Let θ be an approximation of the angle between the coordinate lines measured by

$$\theta = \arctan\left(\frac{y_\xi}{x_\xi}\right) - \arctan\left(\frac{y_\eta}{x_\eta}\right).$$

If the degree of nonorthogonality is limited by the condition

$$|\cot\theta| \le M, \tag{4}$$

then

$$J^2 \ge \frac{1}{M+1}\,|r_\xi|^2 |r_\eta|^2$$

Once it is noted that each term in $T\xi$ has a factor of either x_ξ or y_ξ, it follows that when the quotients in (3) and $\cot\theta$ remain bounded as $|r_\xi| + |r_\eta|$ approaches zero, the order of the difference approximations for first order derivatives is preserved on the curvilinear coordinate system.

The truncation error analysis is considerably more complicated for second order derivatives. The major conclusions will be derived without going into all the technical details. The series expansions for the second order differences are given.

$$f_{\xi\xi} = S^{(1)} + S^{(2)} + \text{HOT}$$

where

$$S^{(1)} = x_{\xi\xi}\frac{\partial f}{\partial x} + y_{\xi\xi}\frac{\partial f}{\partial y} + x_\xi^{\,2}\frac{\partial^2 f}{\partial x^2} + 2x_\xi y_\xi\frac{\partial^2 f}{\partial x\partial y} + y_\xi^{\,2}\frac{\partial^2 f}{\partial y^2}$$

$$S^{(2)} = \frac{1}{4}x_{\xi\xi}^{\,2}\frac{\partial^2 f}{\partial x^2} + \frac{1}{2}x_{\xi\xi}y_{\xi\xi}\frac{\partial^2 f}{\partial x\partial y} + \frac{1}{4}y_{\xi\xi}^{\,2}\frac{\partial^2 f}{\partial y^2}$$

$$+ \frac{2}{3}x_{\xi\xi}x_\xi^{\,2}\frac{\partial^3 f}{\partial x^3} + \frac{1}{2}[x_\xi(x_\xi y_{\xi\xi} + y_\xi x_{\xi\xi}) + \frac{1}{4}x_{\xi\xi}$$

$$(x_\xi y_\xi + x_{\xi\xi}y_{\xi\xi})]\frac{\partial^3 f}{\partial x^2\partial y} + \frac{1}{2}[y_\xi(y_\xi x_{\xi\xi} + x_\xi y_{\xi\xi}) + \frac{1}{4}y_{\xi\xi}$$

$$(x_\xi y_\xi + x_{\xi\xi}y_{\xi\xi})]\frac{\partial^3 f}{\partial x\partial y^2} + \frac{2}{3}y_{\xi\xi}y_\xi^{\,2}\frac{\partial^3 f}{\partial y^3}.$$

The mixed derivative approximation involves diagonal neighbors and it is convenient to introduce the differences

$$f_s = (f(\xi+1,\eta+1) - f(\xi-1,\eta-1))/2\sqrt{2}$$

$$f_t = (f(\xi-1,\eta+1) - f(\xi+1,\eta-1))/2\sqrt{2}$$

$$f_{ss} = (f(\xi+1,\eta+1) + f(\xi-1,\eta-1) - 2f(\xi,\eta))/2$$

$$f_{tt} = (f(\xi-1,\eta+1) + f(\xi+1,\eta-1) - 2f(\xi,\eta))/2.$$

The series expansion has the form

$$f_{\xi\eta} = T^{(1)} + T^{(2)} + \text{HOT}$$

where

$$T^{(1)} = x_{\xi\eta}\frac{\partial f}{\partial x} + y_{\xi\eta}\frac{\partial f}{\partial y} + x_\xi x_\eta \frac{\partial^2 f}{\partial x^2} + (x_\xi y_\eta + x_\eta y_\xi)\frac{\partial^2 f}{\partial x \partial y} + y_\xi y_\eta \frac{\partial^2 f}{\partial y^2}$$

$$T^{(2)} = \frac{1}{2}x_{\xi\eta}(x_{ss} + x_{tt})\frac{\partial^2 f}{\partial x^2} + \frac{1}{2}(x_{\xi\eta}(y_{ss} + y_{tt}) + y_{\xi\eta}(x_{ss} + x_{tt}))\frac{\partial^2 f}{\partial x \partial y} + \frac{1}{2}y_{\xi\eta}(y_{ss} + y_{tt})\frac{\partial^2 f}{\partial y^2}$$

$$+ \frac{2}{3}(x_{ss}x_s^{\ 2} - x_{tt}x_t^{\ 2})\frac{\partial^3 f}{\partial x^3} + \frac{1}{2}[x_s(x_s y_{ss} + y_s x_{ss})$$

$$+ \frac{1}{4}x_{ss}(x_s y_s + x_{ss}y_{ss}) - x_t(x_t y_{tt} + y_t x_{tt})$$

$$- \frac{1}{4}x_{tt}(x_t y_t + x_{tt}y_{tt})]\frac{\partial^3 f}{\partial x^2 \partial y} + \cdots.$$

Once again the terms have been categorized into those used in the derivation of the difference equations, $S^{(1)}$ and $T^{(1)}$, the second and third order terms due to the curvilinear coordinates, $S^{(2)}$ and $T^{(2)}$, and the higher order terms which are representative of truncation error that would be present on a uniform rectangular grid. The two third order terms in $T^{(2)}$ which are omitted would be obtained by interchanging x and y in the two third order terms that are present. Now the differences with respect to s and t can be bounded by differences with respect to ξ and η. Therefore, the only additional conditions, other than those required for first derivatives, in order that the truncation error for second derivattives be the same order as the square of the grid spacing is that the quotients

$$\frac{x_{\xi\eta}}{|r_\xi||r_\eta|}, \quad \frac{y_{\xi\eta}}{|r_\xi||r_\eta|}$$

remain bounded as $|r_\xi| + |r_\eta|$ approaches zero.

In the numerical solution of a partial differential equation, the truncation error may be decreased at a point by adding grid points or moving existing grid points. The technique used to decrease the grid spacing has a significant effect on the local truncation error as the above analysis indicates. This fact is further illustrated in the following one-dimensional examples. Let $x(\xi)$ be an arbitrary fixed grid point. Consider a sequence of grids where the distance between $x^{(n)}(\xi) = x(\xi)$ and its neighbors, $x^{(n)}(\xi-1)$ and $x^{(n)}(\xi+1)$, is decreased by a factor of 2 at each step. Then

$$\frac{x_{\xi\xi}^{(n)}}{(x_\xi^{(n)})^2} = 2\,\frac{x_{\xi\xi}^{(n-1)}}{(x_\xi^{(n-1)})^2} = 2^n\,\frac{x_{\xi\xi}^{(0)}}{(x_\xi^{(0)})^2}$$

and a reduction in the order of the local truncation error would occur unless the original grid was uniform; i.e., $x_{\xi\xi} = 0$. Now consider the case where the grid is defined by a mapping $x(\xi) = f(\zeta)$. Suppose the neighbors of $x(\xi)$ are $x(\xi-1) = f(\zeta-h)$ and $x(\xi+1) = f(\zeta+h)$. Then if $f'(\zeta) \neq 0$ and $f''(\zeta)$ exists

$$\lim_{h\to 0} \frac{x_{\xi\xi}}{x_\xi^2} = \frac{f''(\zeta)}{[f'(\zeta)]^2}$$

is finite. In this case the local truncation error is proportional to the square of the grid spacing or $O(h^2)$. It has been assumed that the function f and the image of $x(\xi)$, which is denoted by ζ, are fixed. This conclusion, that the order is preserved, would not necessarily hold if the function f changed as $h \to 0$.

It was noted above that the degree of skewness in a nonorthogonal coordinate system must be limited to maitain the order of the numerical algorithm. The effect of skewness can be further clarified by noting that

$$J = |r_\xi||r_\eta| \sin\theta.$$

Therefore, for first derivative approximations, the local truncation error varies inversely as the sine of the angle between the coordinate lines. It can also be shown that, for second derivative approximations, an increase in truncation error by a factor of $\sin^{-2}\theta$ is possible. If the skewness is accompanied by large variations in grid spacing, this factor increases to $\sin^{-3}\theta$. The general conclusion is that a moderate degree of skewness has little effect on truncation error. The principal disadvantage in using nonorthogonal coordinates is the added complexity of the difference equations.

Certainly this development does not cover all possible discretizations of a partial differential equation. However, similar conclusions hold for other commonly used difference approximations. In particular, the same conditions for maintaining the order of the difference approximation suffice when second order partial derivatives of the form

$$\frac{\partial}{\partial x}\left(a\frac{\partial f}{\partial x}\right)$$

are approximated by

$$\frac{1}{J}\{y_\eta[\frac{a}{J}(y_\eta f_\xi - y_\xi f_\eta)]_\xi - y_\xi[\frac{a}{J}(y_\eta f_\xi - y_\xi f_\eta)]_\eta\}.$$

Extreme distortions in grid cells can have an especially serious effect in the finite difference analogs of conservation laws. If the partial differential equation

$$\frac{\partial f}{\partial x} + \frac{\partial g}{\partial y} = s \tag{5}$$

is approximated by

$$(fy_\eta - gx_\eta)_\xi + (gx_\xi - fy_\xi)_\eta = Js,$$

then the consistency of the difference equation depends on the difference between

$$[(uv_\eta)_\xi - (uv_\xi)_\eta]/J$$

and

$$[u_\xi v_\eta - u_\eta v_\xi]/J$$

converging to zero where u and v are either of the coordinate variables x and y.

3. FINITE VOLUME EQUATIONS

An alternate method for approximating conservation laws has been widely used on curvilinear coordinate systems. Generally referred to as the finite volume method, it is derived by integrating the differential equation (5) over a grid cell and applying the Gauss Divergence Theorem.

Let C be an arbitrary grid cell with vertices $r(\xi,\eta)$, $r(\xi+1,\eta)$, $r(\xi,\eta+1)$, $r(\xi+1, \eta+1)$, where $r = (x,y)$. Let $f(\xi + \frac{1}{2}, \eta + \frac{1}{2})$ denote the values of the function f at the centroid of C. In the literature, the mean value over the cell is sometimes used rather than the value at the centroid. Since the difference in the two values is the same order as the truncation error, either definition may be assumed. Integrating equation (5) over C gives

$$\int_{\partial C} f\,dy - g\,dx = \int_C s\,dxdy.$$

A third order quadrature may be derived by using the values of f and g at the midpoints of the cell sides. In the usual finite volume formulation, this value on the cell side is approximated by the average of the values on the two cells having the given side in common. Thus the derived difference equation is of the form

$$\sum_{i=1}^{4} (\mu f)_i(\Delta y)_i - (\mu g)_i(\Delta x)_i = As \tag{6}$$

where A is the area of C and, for example,

$$(\mu f)_1 = \frac{1}{2}(f(\xi - \frac{1}{2},\eta + \frac{1}{2}) + f(\xi + \frac{1}{2},\eta + \frac{1}{2}))$$

$$(\Delta x)_1 = x(\xi,\eta) - x(\xi,\eta+1).$$

The effect of the curvilinear coordinate system can be analyzed by considering the difference between μf and the corresponding value of f at the midpoint of the common side. The Taylor series expansion of this difference is, for example,

$$\frac{1}{2}(f(\xi - \frac{1}{2},\eta + \frac{1}{2}) + f(\xi + \frac{1}{2},\eta + \frac{1}{2})) - f(\xi,\eta+\frac{1}{2}) = \tag{7}$$

$$\frac{1}{2}\frac{\partial f}{\partial x}(\xi,\eta + \frac{1}{2})\, x_{\xi\xi}(\xi,\eta + \frac{1}{2}) + \frac{1}{2}\frac{\partial f}{\partial y}(\xi,\eta + \frac{1}{2})\, y_{\xi\xi}(\xi,\eta + \frac{1}{2}) + \text{HOT}$$

where in this case

$$x_{\xi\xi}(\xi,\eta+\tfrac{1}{2}) = x(\xi+\tfrac{1}{2},\eta+\tfrac{1}{2}) + x(\xi-\tfrac{1}{2},\eta+\tfrac{1}{2}) - 2x(\xi,\eta+\tfrac{1}{2}),$$

with

$$x(\xi,\eta+\tfrac{1}{2}) = \tfrac{1}{2}(\,x(\xi,\eta+1) + x(\xi,\eta)).$$

Thus we again see that the order of the quadrature formula is preserved on the curvilinear coordinate system if the quotients in (3) remain bounded as the grid spacing decreases to zero. Here $|r_\xi|$ and $|r_\eta|$ are the distances between the cell centroids. The higher order terms, HOT, in (7) are the same order as the square of the grid spacing and hence do not decrease the third order accuracy that would exist with the midpoint rule.

Due to the simplicity of the equation (5) which was considered, several aspects of the finite volume method have not been mentioned. In practice, s is generally the temporal derivative of some physical quantity. When (6) is solved for s, it is apparent that some lower bound must be placed on A. This is again accomplished by limiting the nonorthogonality and is consistent with results for finite element methods where excessively thin elements are to be avoided. It is also noted that, in most finite volume methods, the equation (6) is implemented in a two-step algorithm to produce second order temporal accuracy.

Basically then, the finite volume methods also require restrictions on the grid to maintain the order of the algorithm. However they can be easier to implement in cases where many rectangular grids are patched together to fill a complicated region. In such cases it is not uncommon for a grid point to have more or less than four neighbors. This causes no problem in deriving finite volume equations, but special difference formulas must be derived when using the finite difference methods as described above. The same comment can be made for cases where triangular grids are produced by singularities in the curvilinear coordinate system.

4. PATCHED COORDINATE SYSTEMS

Thus far we have considered the curvilinear coordinate system at each point to be topologically equivalent to a rectangular cartesian coordinate system in the plane. As has been shown, a loss of accuracy in the numerical algorithm may occur if the grid is severely distorted. Therefore, when the region is too complicated, it may be advisable to partition the region and construct a separate curvilinear coordinate system for each subregion. Most grid generation techniques are flexible enough to permit the smooth continuation of coordinate lines from one subregion into the next. However, at points where the boundaries of several subregions intersect, there may be greater than or less than four neighboring points. An inspection will reveal four grid points which have five neighbors in the grid of Figure 2. Special techniques must be applied to derive

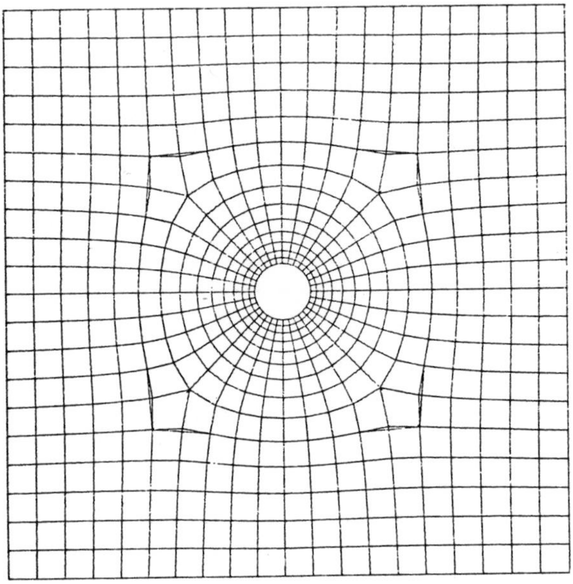

Figure 2.

difference equations at these points. One simple technique would be to select five nearby points and compute a Taylor series truncated after the second order terms. This would give a system of five equations which would be solved to obtain difference approximations for the first and second order derivatives. Unfortunately, the difference equation derived by this method would not resemble the difference equations at other points, which it is assumed would be derived by the usual change of variables formulas. Thus the effect of this differencing technique on numerical properties such as stability and iterative convergence would be uncertain.

A second method of dealing with special points caused by grid patching is more compatible with the usual differencing techniques. Basically it involves selecting nearby points to form a local curvilinear coordinate system. Let $r(\xi,\eta) = (x(\xi,\eta), y(\xi,\eta))$ be a grid point with an excess or deficiency of neighbors. Then four grid points, denoted by $r(\xi \pm 1, \eta)$ and $r(\xi, \eta \pm 1)$, are chosen to define the two coordinate directions through $r(\xi,\eta)$. Four additional points, denoted by $r(\xi \pm 1, \eta \pm 1)$ and $r(\xi \pm 1, \eta \mp 1)$, are chosen from the four quadrants of the new curvilinear coordinate system. The nine points to be used in deriving the difference equation at $r(\xi,\eta)$ have been defined. But the coordinate lines and grid cells may be far from that of a uniform rectangular grid. Therefore, as has been discussed in Section 2, a loss of accuracy is to be expected when the usual finite difference equations are employed. In fact, the local truncation error for the first derivatives will be the same order as the grid spacing. Convergence of second derivative approximations cannot be guaranteed as the grid spacing decreases to zero. Despite this discouraging note, accurate results have been computed using this technique. The inconsistency of the difference approximation for second order equations motivated the search for a higher order approximation. A system of five equations in the five partial derivatives of the function f can be constructed by truncating the Taylor series expansions of the central differences in (1) after the second order terms. The system is written below with the notation of Section 2.

$$\begin{bmatrix} f_\xi \\ f_\eta \\ f_{\xi\xi} \\ f_{\xi\eta} \\ f_{\eta\eta} \end{bmatrix} = \begin{bmatrix} x_\xi & y_\xi & \frac{1}{2}x_\xi x_{\xi\xi} & \frac{1}{2}(x_\xi y_{\xi\xi} + y_\xi x_{\xi\xi}) & \frac{1}{2}y_\xi y_{\xi\xi} \\ x_\eta & y_\eta & \frac{1}{2}x_\eta x_{\eta\eta} & \frac{1}{2}(x_\eta y_{\eta\eta} + y_\eta x_{\eta\eta}) & \frac{1}{2}y_\eta y_{\eta\eta} \\ x_{\xi\xi} & y_{\xi\xi} & x_\xi^2 + \frac{1}{4}x_{\xi\xi}^2 & 2x_\xi y_\xi + \frac{1}{2}x_{\xi\xi}y_{\xi\xi} & y_\xi^2 + \frac{1}{4}y_{\xi\xi}^2 \\ x_{\xi\eta} & y_{\xi\eta} & x_\xi x_\eta + \frac{1}{2}x_{\xi\eta}(x_{ss}+x_{tt}) & x_\xi y_\eta + x_\eta y_\xi + x_{\xi\eta}(y_{ss}+y_{tt}) + y_{\xi\eta}(x_{ss}+x_{tt}) & y_\xi y_\eta + \frac{1}{2}y_{\xi\eta}(y_{ss}+y_{tt}) \\ x_{\eta\eta} & y_{\eta\eta} & x_\eta^2 + \frac{1}{4}x_{\eta\eta}^2 & 2x_\eta y_\eta + \frac{1}{2}x_{\eta\eta}y_{\eta\eta} & y_\eta^2 + \frac{1}{4}y_{\eta\eta}^2 \end{bmatrix} \begin{bmatrix} \frac{\partial f}{\partial x} \\ \frac{\partial f}{\partial y} \\ \frac{\partial^2 f}{\partial x^2} \\ \frac{\partial^2 f}{\partial x \partial y} \\ \frac{\partial^2 f}{\partial y^2} \end{bmatrix} \qquad (8)$$

For a uniform rectangular grid, the usual difference approximations are produced. Although it cannot be guaranteed that the coefficient matrix is well-conditioned or at least nonsingular, this is suggested by the fact that the system (8) is a perturbation of the nonsingular system of equations which produces the usual difference equations. The later system has a coefficient matrix whose determinant is J^3. This technique generates a nine-point difference equation using the same differences on the local coordinate system that are used at the other points of the grid. The local truncation error for first derivative approximations is the same order as the square of the grid spacing, whereas, the local truncation error for second derivative approximations is the order of the grid spacing. This result is valid regardless of the coordinate line spacing or curvature. It is only assumed that the coefficient matrix is not ill-conditioned.

While it is possible to generate consistent difference equations at special points encountered in grid patching, it should be noted that loss of accuracy is possible. The condition that grid lines pass smoothly from one region into the next also places a restriction on the number and location of grid points in each subregion. Coordinate lines which are discontinuous, or have discontinuous slopes, at subregion boundaries can be used, but this further complicates the problem of deriving accurate difference equations.

5. OVERLAPPING COORDINATE SYSTEMS

When dealing with complicated regions, especially multiply connected regions, there may be portions of the boundary where a curvilinear coordinate system can be easily constructed. For example, consider the region between a rectangle and a circle. A polar coordinate system is the obvious choice near the circle while a cartesian coordinate system would be best near the rectangle. Each of these coordinate systems can be extended into the region until they overlap and form a covering of the region by grid cells as illustrated in Figure 3. In general, there may be several overlapping grid systems used to cover a particular region. The

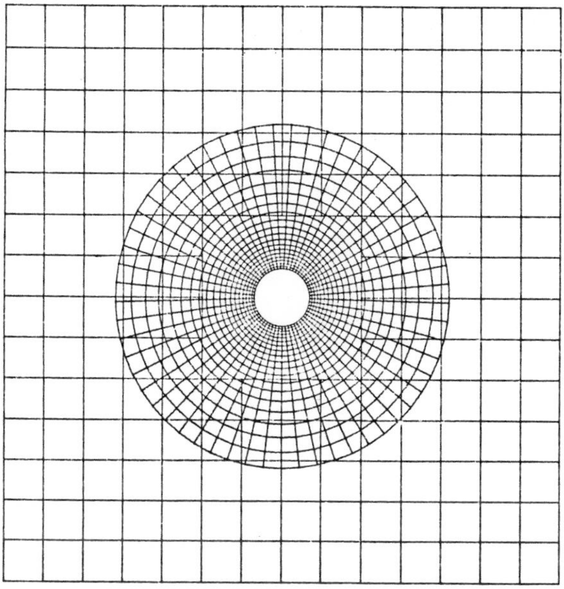

Figure 3.

difference equations must couple the solution values on the various grid systems. This transmission of information is most frequently accomplished by interpolation at those grid boundary points which lie in the interior of the region. Several interpolation procedures will be examined along with their impact on finite difference methods.

Let $G^{(1)}$ be a grid with boundary point r_0 which is contained in some grid cell of the grid $G^{(2)}$. First, the general interpolation formula

$$f(r_0) = \sum_{j=1}^{k} \alpha_j f(r_j) \tag{9}$$

will be considered where r_j denotes a point in $G^{(2)}$. When the value of f at r_0 is replaced in a difference equation by the interpolated value, a new difference equation results which may have a different local truncation error. Conditions on the coefficients α_j which will preserve the order of the difference equation can be derived by examining the effect on a Taylor series at an interior neighbor r of r_0. If the value of f at r_0 is computed from the values at r_j by (9), then

$$f(r_0) = \sum_{j=1}^{k} \alpha_j\ f(r) + \sum_{j=1}^{k} \alpha_j (x_j - x) \frac{\partial f}{\partial x}(r)$$

$$+ \sum_{j=1}^{k} \alpha_j (y_j - y) \frac{\partial f}{\partial y}(r)$$

$$+ \sum_{j=1}^{k} \alpha_j (x_j - x)^2 \frac{\partial^2 f}{\partial x^2}(r)$$

$$+ 2\sum_{j=1}^{k} \alpha_j (x_j - x)(y_j - y) \frac{\partial^2 f}{\partial x \partial y}(r)$$

$$+ \sum_{j=1}^{k} \alpha_j (y_j - y)^2 \frac{\partial^2 f}{\partial y^2}(r) + \cdots .$$

This series coincides with the actual Taylor series, computed at r_0, through first order terms if

$$\sum_{j=1}^{k} \alpha_j = 1, \quad \sum_{j=1}^{k} \alpha_j x_j = x_0, \quad \sum_{j=1}^{k} \alpha_j y_j = y_0, \tag{10}$$

and through second order terms if, in addition,

$$\sum_{j=1}^{k} \alpha_j (x_j)^2 = (x_0)^2, \quad \sum_{j=1}^{k} \alpha_j x_j y_j = x_0 y_0, \quad \sum_{j=1}^{k} \alpha_j (y_j)^2 = (y_0)^2. \tag{11}$$

It will be assumed that the conditions, indicated in Section 2, for preserving the order of first and second order difference equations hold for the individual grids $G^{(1)}$ and $G^{(2)}$. In that case, the local truncation error for first derivative approximations would be the same order as the grid spacing if (10) holds and the same order as the square of the grid spacing if both (10) and (11) hold. The order of the local truncation error for second order differences would be the same as the grid spacing if both (10) and (11) hold. The order would be the square of the grid spacing if an additional condition equating the coefficients of third order terms in the series was imposed. The condition which equates the p th order coefficients can be written as

$$\sum_{j=1}^{k} \alpha_j [x_j]^m [y_j]^{n-m} = [x_0]^m [y_0]^{n-m}, \tag{12}$$

$n + m = p$, $m = 0, 1, \cdots, p$.

Implicit schemes tend to be difficult to implement on regions which use several curvilinear coordinate systems. Therefore, most currently used algorithms involve the iterative solution of elliptic equations or the explicit solution of parabolic or hyperbolic equations. This naturally leads one to question the effect of the interpolation equation on iterative convergence in the first case and on stability in the later case. No detailed analysis will be given here, but a few obvious comments are worth noting. Diagonal dominance of the coefficient matrix of the difference equations is a sufficient, although not necessary, condition for the convergence of many iterative methods. A system of diagonally dominant difference equations will remain diagonally dominant when the interpolation equations (9) are appended if

$$\sum_{j=1}^{k} |\alpha_j| \le 1. \tag{13}$$

However, when this condition is considered with the first equation in (10), which is necessary for consistency, it follows that diagonal dominance will be preserved whenever

$$\alpha_j \ge 0, \ j = 1, 2, \cdots, k. \tag{14}$$

The stability properties of (9) can also be observed. Suppose the value at r_0 is computed at $t = (n+1)\Delta t$ by

$$f(r_0,(n+1)\Delta t) = \sum_{j=1}^{k} \alpha_j f(r_j, n\Delta t). \tag{15}$$

The von Neumann stability analysis is based on the behavior of an exponential solution of the form

$$f(x,y,t) = \exp(\lambda t)\ \exp(i\mu x)\ \exp(i\nu y).$$

Substituting in (15) gives the following

$$\exp(\lambda\Delta t) = \sum_{j=1}^{k} \alpha_j \ \exp(i\mu\Delta x_j)\ \exp(i\nu\Delta y_j),$$

where $\Delta x_j = x_j - x_0$ and $\Delta y_j = y_j - y_0$. For real μ and ν, the exponential solution will remain bounded as $n \to \infty$ provided (13) holds. Therefore, whenever (14) holds along with (10), the interpolation equations impose no additional stability restriction on the numerical algorithm.

Several different interpolation schemes will be reviewed in light of the above remarks. The first scheme is based on the approximation of a linear Taylor polynomial. For each boundary point r_0 of $G^{(1)}$, select a nearby point r_1 of $G^{(2)}$. The neighbors of r_1 in $G^{(2)}$ are indexed as in Figure 4 so that

$$f_\xi(r_1) = (f(r_3) - f(r_2))/2$$

$$f_\eta(r_1) = (f(r_5) - f(r_4))/2.$$

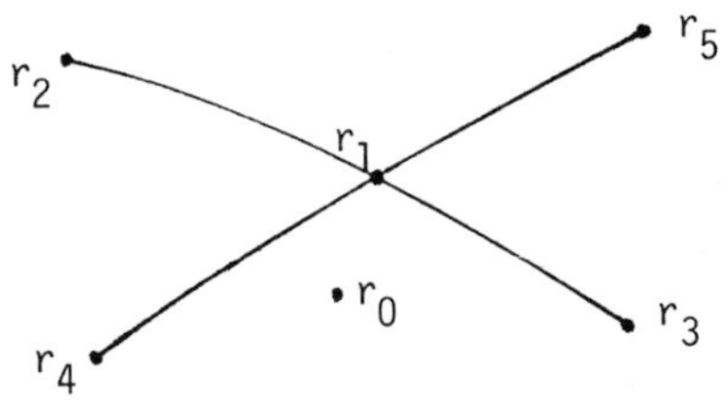

Figure 4.

These differences can be used to approximate the partial derivatives with respect to x and y. If the approximations are substituted in a Taylor series, truncated after the linear terms, then the resulting interpolation formula can be written

$$f(r_0) = \sum_{j=1}^{k} \alpha_j f(r_j),$$

where

$$\alpha_1 = 1$$

$$\alpha_2 = -\alpha_3 = ((x_0 - x_1)\, y_\eta - (y_0 - y_1)x_\eta)/2J$$

$$\alpha_4 = -\alpha_5 = ((y_0 - y_1)\, x_\xi - (x_0 - x_1)y_\xi)/2J.$$

In this case, it easily observed that equations (10) are valid, but (14) does not generally hold. Second degree Taylor polynomials have been used, but will not be considered here. It is doubtful that they would give better results since the approximation of second derivatives from the numerical solution may be very inaccurate. There are other interpolation schemes for which both (10) and (14) hold and these will be investigated next.

Let r_0 belong to a grid cell C of $G^{(2)}$ with vertices which will be denoted by r_j, j = 1, 2, 3, 4, as illustrated in Figure 5. There exists

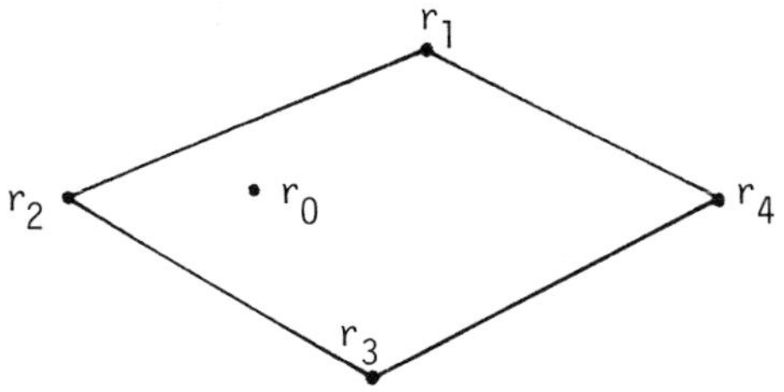

Figure 5.

a unique bilinear mapping of the unit square onto the cell C. The mapping can be given explicitly by

$$r = (1-s)(1-t)\, r_1 + s(1-t)\, r_2 + t(1-s)\, r_4 + st\, r_3,$$

where $0 \leq s, t \leq 1$. If we set $r = r_0$, then the system of two equations, in terms of the x and y coordinates, can be solved to determine the solution $s = s_0$, $t = t_0$. If f is also assumed to be a bilinear function of s and t, then

$$f(r_0) = (1-s_0)(1-t_0)\, f(r_1) + s_0(1-t_0)\, f(r_2) + t_0(1-s_0)\, f(r_4) + s_0 t_0\, f(r_3). \tag{16}$$

It is immediately evident that this interpolation formula satisfies both (10) and (14). Although equations (11) would not be satisfied, this method can be modified to give higher order interpolants. Basically it involves constructing a bicubic mapping of a square grid in the st-plane onto the union of the nine cells consisting of C and the eight cells having an edge or vertex in common with C. This mapping can be expressed in terms of cubic Lagrange interpolating polynominals. The only additional difficulty is that the bicubic equations which determine s_0 and t_0 would now have to be solved numerically whereas the values of s_0 and t_0 in (16) can be computed exactly. By construction, the coefficients for the bicubic interpolation will satisfy (12) for $p = 0, 1, 2, 3$. Therefore the interpolation scheme would not increase the local truncation error. However, (14) would not be valid.

Interpolation on triangular regions is very popular in finite element analysis. Some of those ideas can be adapted to the present problem. Suppose each quadrilateral cell is divided into two triangular cells. Then r_0 will belong to a triangular cell with vertices r_1, r_2, r_3. Note that this would be the case in Figure 5 if the quadrilateral is partitioned by the diagonal from r_1 to r_3. Now the three equations in (10) determine the coefficients for the interpolation formula (9) which coincides with linear interpolation on the triangular cell. As long as r_0 is an interior or boundary point of the cell, the condition (14) will be satisfied. The accuracy of the interpolation formula can be increased by increasing the

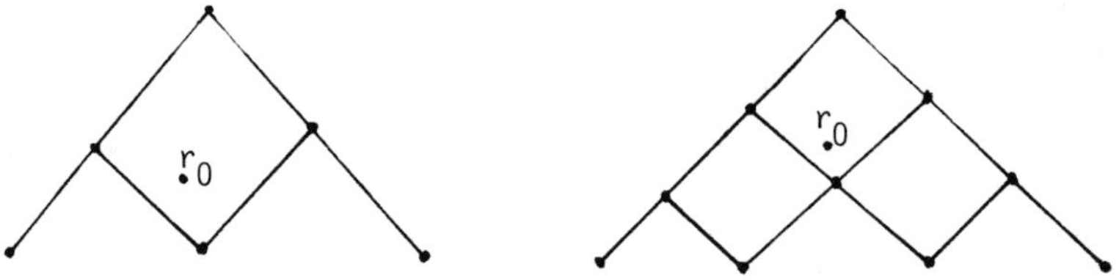

Figure 6.

number of interpolation points. Figure 6 indicates the grid points which could be used for quadratic and cubic interpolation. In each case, the coefficients in (9) can be calculated from the general equations in (12). The quadratic interpolant uses six coefficients obtained by setting $p = 0, 1, 2$ and the ten coefficients for the cubic interpolant are found by solving (12) with $p = 0, 1, 2, 3$. A few cautionary notes are in order. The linear interpolation polynomial always exists, but for severely distorted grids, the system in (12) may be singular and the interpolation polynomial may not exist in the quadratic and cubic case. Condition (14) is also not satisfied in the quadratic and cubic case.

Several interpolation schemes have been presented for use on overlapping coordinate systems. This does not include all techniques which are presently in use. In particular, we have not considered methods which interpolate normal derivatives at the boundary points of each grid. There is no reason why this analysis cannot be extended to cover that case. We would only need consider the formula (9) with some of the r_i in $G^{(1)}$ and the remaining r_i in $G^{(2)}$.

6. CONCLUSIONS.

The procedure for selecting a curvilinear coordinate system must necessarily involve a balance of certain requirements. The rate of change in coordinate line spacing and degree of skewness should be limited so that the formal accuracy of the difference equation is maintained. On the other hand, efficient use of grid points mandates the clustering of points in regions where the derivatives of the theoretical solution are large. If one must use a highly distorted coordinate system or is faced with the prospect of connecting many separate curvilinear coordinate systems in different subregions, it is generally possible to derive consistent difference approximations. While higher order approximations may exist, their use may not be necessary or advisable. The grid spacing is only one factor in the local truncation error. The other factor is the theoretical solution of the partial differential equation. In a region where all derivatives of the solution are negligible, the local truncation error will be small regardless of the order. Consequently, when solving fluid flow problems, the accuracy of the numerical algorithm is most likely to be maintained if irregularities in the grid can be confined to regions of free stream flow. There are multitudes of examples where the use of higher order methods produce inferior results for one reason or the other. In connection with the use of interpolation for overlapping coordinate systems, it should be recalled that Lagrange interpolating polynomials may be highly oscillatory.

The analysis of error for nonlinear systems of partial differential equations solved numerically on large computational grids can never be precise. However the quality of a numerical solution can often be judged by examining the grid and the point-to-point variation in the numerical solution at the grid points and possibly by recomputing the solution on a properly refined grid.

REFERENCES

1. Atta, E. H., Component-adaptive grid embedding, in Numerical Grid Generation Techniques (R. E. Smith, ed.), NASA CP 2166, NASA Langley Research Center, 1980, 157-174.
2. ______________ and J. Vadyak, A grid interfacing zonal algorithm for three-dimensional transonic flows about aircraft configurations, AIAA Paper No. 82-1017, AIAA/ASME 3rd Joint Thermophysics, Fluids, Plasma and Heat Transfer Conference, St. Louis, June 1982.
3. Hoffman, J. D., Relationship between the truncation error of centered finite-difference approximation on uniform and nonuniform meshes, J. Comput. Phys., 46(1982), 469-474.
4. Lee, K. D., M. Huang, N. J. Yu, and P. E. Rubbert, Grid generation for general three-dimensional configurations, in Numerical Grid Generation Techniques (R. E. Smith, ed.), NASA CP 2166, NASA Langley Research Center, 1980, 367-376.
5. Roberts, A., Automatic topology generation and generalized B-spline mapping, in Numerical Grid Generation (J. F. Thompson, ed.) Elsevier/North-Holland, 1982.
6. Starius, G., Composite mesh difference methods for elliptic boundary value problems, Numer. Math., 28(1977), 243-258.
7. Thompson, J. F., Z. U. A. Warsi, and C. W. Mastin, Boundary-fitted coordinate systems for numerical solution of partial differential equations - a review, J. Comput. Phys., 47(1982), 1-108.
8. Vinokur, M., On one-dimensional stretch-functions for finite-difference calculations, NASA CR 3313, Ames Research Center, 1980.

The author was supported by NASA Grant NSG 1577, Langley Research Center.

Department of Mathematics & Statistics
Mississippi State University
Mississippi State, MS 39762

NUMERICAL AERODYNAMIC SIMULATION (NAS)

V. L. Peterson, W. F. Ballhaus, Jr., and F. R. Bailey

1. INTRODUCTION

Computational aerodynamics has emerged in the last decade as a powerful tool for the design of all types of aerospace vehicles, but it is one of several technical disciplines whose development is being restrained by the lack of suitably powerful computational systems. The equations of fluid flow are well-known, but the methods for solving them without approximation require unrealistically large amounts of computation for all but the most elementary flow situations. Thus, the aerodynamicist must resort to the use of approximate forms of the equations to treat complete aircraft geometries. In fact, for three-dimensional configurations, only the equations for inviscid flow can be solved in a practical amount of time with currently available computers. This limitation on the treatment of fluid-flow physics severely restricts a designer's ability to analyze aerodynamic phenomena and the resulting effects on aircraft performance.

The Numerical Aerodynamic Simulation (NAS) Program is designed to provide a leading-edge computational capability to the aerospace community. It was recognized early in the program that, in addition to more advanced computers, the entire computational process ranging from problem formulation to publication of results needed to be improved to realize the full impact of computational aerodynamics. Therefore, the NAS Program has been structured to focus on the development of a

ISBN 0-12-546080-5

complete system that can be upgraded periodically with minimum impact on the user and on the inventory of applications software. The implementation phase of the program is now under way. It is based upon nearly 8 yr of study and should culminate in an initial operational capability before 1986.

The objective of this paper is fivefold: (1) to discuss the factors motivating the NAS program, (2) to provide a history of the activity, (3) to describe each of the elements of the processing-system network, (4) to outline the proposed allocation of time to users of the facility, and (5) to describe some of the candidate problems being considered for the first benchmark codes.

2. MOTIVATING FACTORS

The discipline of aerodynamics plays an important role in the design of aerospace vehicles. Beginning nearly 200 yr before the Wright Brothers' historic powered flight in 1903 and continuing until the 1950s when electronic digital computers first became widely available, aerodynamic research and development was based largely on experimentation. This was augmented with some guidance from closed-form analytical solutions of highly approximate forms of the now well-known governing equations of fluid flow, after they were first published by Claude Luis M. H. Navier and Sir George G. Stokes in 1823 and 1845, respectively. Since the 1950s, advancements in computers have offered the possibility of obtaining solutions for successively refined approximations of the Navier-Stokes equations. These solutions have provided an increasingly important complement to experiments being conducted in wind tunnels and in flight. Today, the two principal tools of the aerodynamicist are wind tunnels and computers, and the balance between the use of these tools is beginning to shift in the direction of the computer.

Both technical and economic factors are driving the aerodynamicists toward computers, even though achievements in aeronautics continue to be made largely on the basis of experiment. As the speed and complexity of aircraft have increased, the inherent limitations of wind tunnels as the primary tool for developing and verifying the design of new aircraft and aerospace vehicles have become increasingly apparent. For example, the walls in wind tunnels and the supports used to hold models severely restrict the accuracy of measurements obtained at near-sonic speeds. Aeroelastic distortions, which are known to always be present in flight, cannot be accurately simulated with subscale models tested in wind tunnels. Furthermore, limitations on temperature and on the working fluid restrict the ability of the wind tunnel to simulate flight entry into Earth's atmosphere and entry of probes into other planetary atmospheres. These limitations did not seriously affect the usefulness of wind tunnels during the era of simpler, low-speed flight. The increased complexity and broadened performance envelopes of aircraft have increased the number of wind-tunnel hours expended in the development of a new aircraft, and the cost per hour of testing has been increasing substantially at the same time. On the other hand, the cost of numerically simulating a given flow is decreasing at a rapid rate because of improvements in algorithms and computer cost effectiveness.

Even with improvements in wind tunnels and testing procedures over the years and with the use of the more sophisticated theoretical approaches made possible by computers, the available simulation capability has been inadequate to uncover design problems with a number of aircraft before full-scale prototypes were tested in flight. Some examples of these problems and their consequences are summarized for several aircraft in Table 1. All of these aircraft were developed a

Table 1. Examples of design problems resulting from inadequate simulation capability

AIRCRAFT	PROBLEMS DISCOVERED IN FLIGHT TEST	CONSEQUENCE
MILITARY TRANSPORT-1	INCORRECTLY PREDICTED WING FLOW	COMPROMISED PERFORMANCE, COSTLY MODIFICATIONS
MILITARY TRANSPORT-2	INCORRECTLY PREDICTED DRAG-RISE MACH NUMBER	REDUCED WING FATIGUE LIFE
FIGHTER/ BOMBER	INCORRECTLY PREDICTED TRANSONIC AIRFRAME DRAG	COSTLY MODIFICATIONS
BOMBER-A BOMBER-B	INCORRECTLY PREDICTED TRANSONIC PERFORMANCE	REDUCED AIRCRAFT EFFECTIVENESS
FIGHTER-1 FIGHTER-2	INCORRECTLY PREDICTED TRANSONIC DRAG	REDUCED PERFORMANCE
2 CIVIL TRANSPORTS	INCORRECTLY PREDICTED NACELLE-WING INTERFERENCE	REDESIGN REQUIRED

number of years ago, before computers were sufficiently powerful to have a significant impact on aerodynamic design. Some of the problems could have been avoided using today's level of computational capability, whereas others could not have been uncovered without computers considerably larger than those currently available. Successful design of next-generation aircraft will be even more difficult without improvements in simulation capabilities. Therefore, there are compelling reasons to advance the state of the art in computational aerodynamics. This would require more powerful computers and a continuing vigorous program to improve computational methodology.

The rapid growth in the capability of computational methods to treat problems of practical importance began in the early 1970s when computers became large enough to treat three-dimensional transonic flows requiring the use of nonlinear forms of the governing equations. The first transonic solutions for a lifting airfoil with an embedded shock wave, which could only be solved with a computer, appeared in the literature in 1970 [1]. One of the first solutions for a wing in transonic flow was obtained in 1972; it took 18 hr of computation on an IBM 360 machine. Just 8 yr later, the results presented in Fig. 1 for the flow about a complete KC 135 aircraft with winglets at Mach 0.78 were published [2]. Chordwise pressure coefficients at selected wing stations are shown to compare favorably with measurements for this situation where neglected viscous effects are not dominant. Remarkably, the calculations required only about 15 min of time on a CDC 7600 computer.

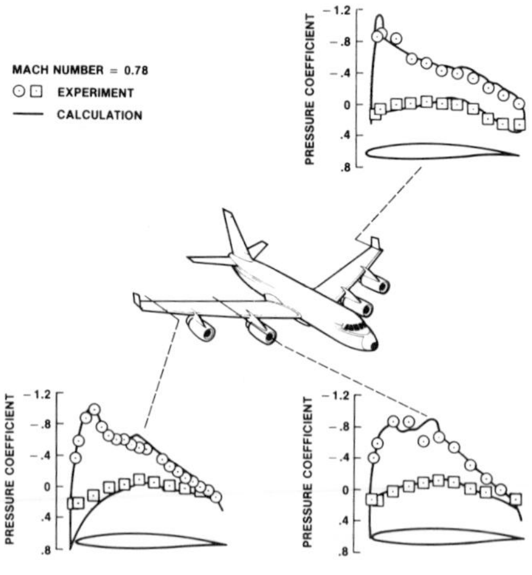

Fig. 1. Example Application of Nonlinear Inviscid Equations to Simulate Flow About a Modified Boeing KC-135 Airplane.

Computers of the 1970s were large enough to solve, in a practical amount of time, inviscid forms of the governing equations that provide satisfactory results for relatively complex but aerodynamically clean configurations operating at or near cruise conditions. However, many of the more crucial design situations are difficult, if not impossible, to treat without including the effects of viscosity in a fully coupled fashion. Examples of these are inlet, engine, and exhaust flows; airframe propulsion system integration; stall and buffet; vortex enhanced lift; maneuvering loads; and performance near performance boundaries. These are also very difficult to treat effectively in wind-tunnel tests.

The Reynolds-averaged form of the Navier-Stokes equations is suitable for treating most of the design problems that are dominated by viscous effects. Revolutionary advances in this technology were made in the 1970s, beginning with the investigation of shock-wave interaction with a laminar boundary layer in 1971 [3] and continuing into the 1980s with the treatment of high-Reynolds-number airfoils, bodies of revolution, and corner flows. This work has been limited to so-called building-block flows, however, because of the large amounts of computer time required for all but the most basic aircraft components. One to 20 hr of computation are needed for many applications, even with the use of the Class VI machines. An example of the application of viscous-flow technology to a turbulent supersonic flow over an axisymmetric conical afterbody containing a centered propulsive exhaust flow [4] is presented in Fig. 2. Computed density contours in the vicinity of the boattail region and the exhaust plume show excellent qualitative agreement with features of the flow made visible with schlieren photography in an experimental investigation reported for the same flow conditions [5].

A conclusion that may be drawn from the aforementioned examples is that problems involving complex geometries can be treated with simple physics and those involving simple geometries can be treated with complex physics, but more powerful computers with larger memories are required to solve problems involving both complex geometries and complex physics. The amount of computer power required to advance the capability of computational aerodynamics to treat increasingly complex aerodynamic flows is indicated by the data in Table 2. Solutions

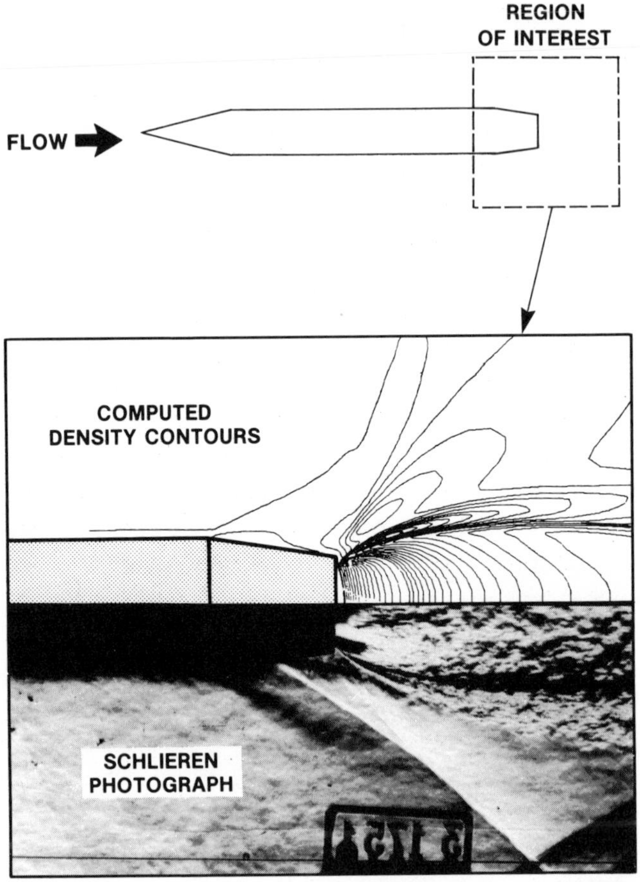

Fig. 2. Comparison of Numerically Simulated and Observed Turbulent Flow over an Axisymmetric Conical Afterbody Containing a Centered Propulsive Jet: Stream Mach Number = 2, Jet Mach Number = 2.5, Jet Static Pressure — Nine Times Free-stream Static Pressure.

of increasingly refined approximations to the full Navier-Stokes equations provide additional information of value to the designer, but they also require increasing refinements to the computational mesh to capture the additional detail in the flow physics. This places an increasing burden on the computer used to solve the equations. The computer requirements given in the table are based on obtaining solutions to problems involving the number of grid points shown in 10 to 15 min with algorithms expected to be available in 1985. Class VI computers are adequate for the inviscid-flow technology, but

Table 2. Computer requirements for treating increasingly complex forms of the governing equations

APPROXIMATION	CAPABILITY	GRID POINTS REQUIRED	COMPUTER REQUIREMENT
LINEARIZED INVISCID	SUBSONIC/SUPERSONIC PRESSURE LOADS VORTEX DRAG	3×10^3 PANELS	1/10 CLASS VI
NONLINEAR INVISCID	ABOVE PLUS: TRANSONIC PRESSURE LOADS WAVE DRAG	10^5	CLASS VI
REYNOLDS AVERAGED NAVIER-STOKES	ABOVE PLUS: SEPARATION/REATTACHMENT STALL/BUFFET/FLUTTER TOTAL DRAG	10^7	30 x CLASS VI
LARGE EDDY SIMULATION	ABOVE PLUS: TURBULENCE STRUCTURE AERODYNAMIC NOISE	10^9	3000 x CLASS VI
FULL NAVIER-STOKES	ABOVE PLUS: LAMINAR/TURBULENT TRANSITION TURBULENCE DISSIPATION	10^{12} TO 10^{15}	3 MILLION TO 3 BILLION CLASS VI

they fall far short of meeting the requirements for problems dominated by strongly coupled viscous effects. The mext major step forward for the discipline of computational aerodynamics can be taken when computers about 30 times more powerful than the Class VI machines become available.

3. PROGRAM HISTORY

The Numerical Aerodynamic Simulation Program had its genesis in 1975 at the NASA Ames Research Center. A small group of people associated with the computational fluid dynamics program proposed the development of a special-purpose Navier-Stokes Processing Facility which would have several orders of magnitude more speed than the ILLIAC IV computer for solving the equations of fluid physics. The proposal called for the central processor to have a minimum effective speed of one billion floating point operations per second when operating on the three-dimensional Reynolds-averaged Navier-Stokes equations and have performance comparable to the best general-purpose computers when used for processing the equations of other scientific disciplines. Its working memory had to accommodate a problem data base of at least 31×10^6 64-bit words. To keep development risks low, the goal of the project was to assemble existing technologies into a specialized architecture rather than to develop new electronic components. Finally, the machine had to be user-oriented, easy to program, and capable of detecting systematic errors when they occurred. The proposal was endorsed in principle by NASA management in Nov. 1975 and the name of the project was changed to the Computational Aerodynamic Design Facility.

3.1 Computational Aerodynamic Design Facility Project

The first formal exposure of NASA's objectives occurred in October 1976 when proposals were requested from industry to "perform analysis and definition of candidate configurations for a computational facility in order to arrive at the best match between computational aerodynamic solution methods and processor system design." These analyses were to be directed toward the selection, preliminary design, and evaluation of candidate system configurations that would be best suited to the solution of given aerodynamic flow models. Design requirements that were established for this study included (1) the capability to complete selected numerical solutions of the Navier-Stokes equations for grid sizes ranging from 5×10^5 to 1×10^6 points and wall clock times (exclusive of input-data preparation and output-data analysis) ranging from 5 to 15 min; (2) a working memory of 40×10^6 words; (3) an archival store of at least 10×10^9 words; and (4) 120 hr/wk of availability to the users.

Two parallel contracts were awarded in Feb. 1977 to develop preliminary designs for the most promising configurations and to develop performance estimates, risk analyses, and preliminary implementation cost and schedule estimates for each of these designs. During these initial studies, which lasted about 12 mo, it became apparent that the overall approach to developing the facility was sound and that performance goals could be reached with new architectural concepts and proven electronic components.

3.2 Numerical Aerodynamic Simulation Facility Project

When it was recognized that the facility would be used primarily for computational research rather than for actual aircraft design, the name was changed to the Numerical

Aerodynamic Simulation Facility shortly after the first study contracts were awarded. It also became apparent that a computational resource of this magnitude would be a valuable tool for the solution of complex problems in other technical areas of interest such as meteorology, chemistry, astrophysics, and structures. Plans were made at that time to survey the aerodynamics community and technical universities to ascertain the level of interest in the projected facility and to ensure that design goals were consistent with the needs of the projected users of the facility.

Before the conclusion of the first round of contracted efforts, the need for further studies with greater emphasis on a computer suitable for a broader range of disciplines was recognized. Accordingly, 12-mo follow-on feasibility study contracts were awarded in March 1978. The results of these efforts were expected to provide data of sufficient accuracy to permit formulation of a definitive plan for development of the facility. Several events occurred during the period of these studies which resulted in some revisions to the basic performance specifications and a deeper involvement of the user community in the project activities.

The discipline of computational aerodynamics had matured significantly in the 3 yr since the project was first conceived. New numerical methods were developed and existing methods were refined. This led to the realization that if the size of the on-line or working memory was increased to 240×10^6 words, the facility could be used not only to estimate the performance of relatively complete aircraft configurations, but also to serve as an effective tool to study the physics of turbulent flows, a subject eluding researchers for more than 80 yr. A corresponding increase in the off-line file storage from 10×10^9 to approximately 100×10^9 words was required to accommodate the larger data sets.

A User Steering Group was formed in 1978 to provide a channel for the dissemination of information regarding project status, a forum for user-oriented issues needing discussion, and a sounding board by which the project office could obtain feedback from future user organizations. Examples of user-oriented issues of interest are (1) selection of user languages, (2) management policy, (3) equipment required for remote access, and (4) data protection. The User Steering Group was composed of representatives of the aerospace industry, universities, and other government agencies. It is still active although its name was changed recently to the User Interface Group to more accurately reflect its current role.

The feasibility studies were completed in the spring of 1979. Each study produced a refined baseline configuration, a functional design, and rough estimates of cost and schedule. Both studies concluded that ~5 yr would be required to complete the detailed design and to develop, integrate, and test the facility. While preparations were being made to continue the contracted development process, the name of the project was changed once again to the Numerical Aerodynamic Simulator (NAS) Project.

3.3 Numerical Aerodynamic Simulator Project

A detailed plan for the design-definition phase of activity was prepared during the winter of 1979 by the NAS Project Office, which was established at Ames earlier in the year. This included refining (1) the specifications for the computing engine, (2) the support processing system, and (3) the collection of other peripherals, including intelligent terminals, graphical display devices, and data communication interfaces to local and remote users. Two 40-wk parallel design-definition contracts were awarded in Sept. 1980. Upon their completion in July 1981, the contractors were awarded follow-on contracts related to further design definition. These were concluded in April 1982 when the proposals for the detailed design, development, and construction were submitted by the contractors for evaluation.

After an initial evaluation of the proposals, the decision was made to abandon the previously defined approach to meeting the requirements of computational aerodynamics and to chart a new course. Several factors contributed to this decision. First, the application and essential importance of computational aerodynamics had grown significantly since the mid-1970s. Thus, it was deemed important to establish and sustain a leading-edge computational capability as an essential step toward maintaining the nation's leadership in aeronautics. To achieve this goal the NAS Project was to be restructured as an on-going NAS Program in which significant advances in high-speed computer technology would be continuously incorporated as they become available.

Second, the supercomputer environment had changed since the inception of the NAS activity in the 1970s. Increased interest in supercomputing, along with increased foreign competition, changed the environment to the extent that it no longer appeared necessary to directly subsidize the development of the next generation of scientific computers. Rather, the current environment permits a more systematic, evolutionary approach toward developing and maintaining an advanced NAS computational capability.

Finally, it was recognized that advancement in the state of the art of supercomputers must be coupled with advanced system networks and software architectures. This is necessary to accommodate successive generations of supercomputers from different vendors and to provide the capabilities needed to enhance productivity for the user. This leads to a strategy that minimizes the dependence of the entire system on single vendors and leads to the establishment of a strong in-house technical capability to direct the initial and ongoing development efforts.

3.4 Numerical Aerodynamic Simulation Program

A plan for the redefined program was approved in Feb. 1983. It includes (1) the design, implementation, testing and integration of an initial operating configuration of the NAS Processing System Network; (2) the systematic and evolutionary incorporation of advanced computer-system technologies to maintain a leading-edge performance capability; and (3) the management and operation of the complex. The principal goals of the program are summarized as follows.

General goals

Optimize the computational process from problem formulation to publication of results.

Specific goals

- 1985
  - Speed — 250 MFLOPS, sustained
  - Memory — at least 64×10^6 64-bit words
  - Users — local and remote
- 1987
  - Speed — 1 GFLOPS, sustained
  - Memory — 256×10^6 64-bit words
  - Users — support at least 100 simultaneously on a time-sharing interactive basis
  - Operating system and network — capable of accommodating a multivendor hardware environment
- Beyond 1987
  - Continue to expand capability

4. NAS PROCESSING SYSTEM NETWORK

The planned NAS Processing System Network (NPSN) will be a large-scale, distributed resource computer network at Ames Research Center. This network will provide the full end-to-end capabilities needed to support computational aerodynamics, will span the performance range from supercomputers to microprocessor-based work stations, and will offer functional capabilities ranging from "number-crunching" interactive aerodynamic-flow-model solutions to real-time graphical-output-display manipulation. The NPSN resources will be made available to a nationwide community of users via interfaces to landline and satellite data communications links.

The NAS program is structured to accommodate the continuing development of the NPSN as a leading-edge computer-system resource for computational aerodynamics. This development process is dependent upon the acquisition and integration of the most advanced supercomputers industry can provide that are consistent with computational aerodynamics requirements. Figure 3 illustrates the continuing development of the NPSN functional and performance capabilities while successively introducing advanced high-speed processors into the network. The introduction of each new high-speed processor involves an integration phase in which new software and interfaces are

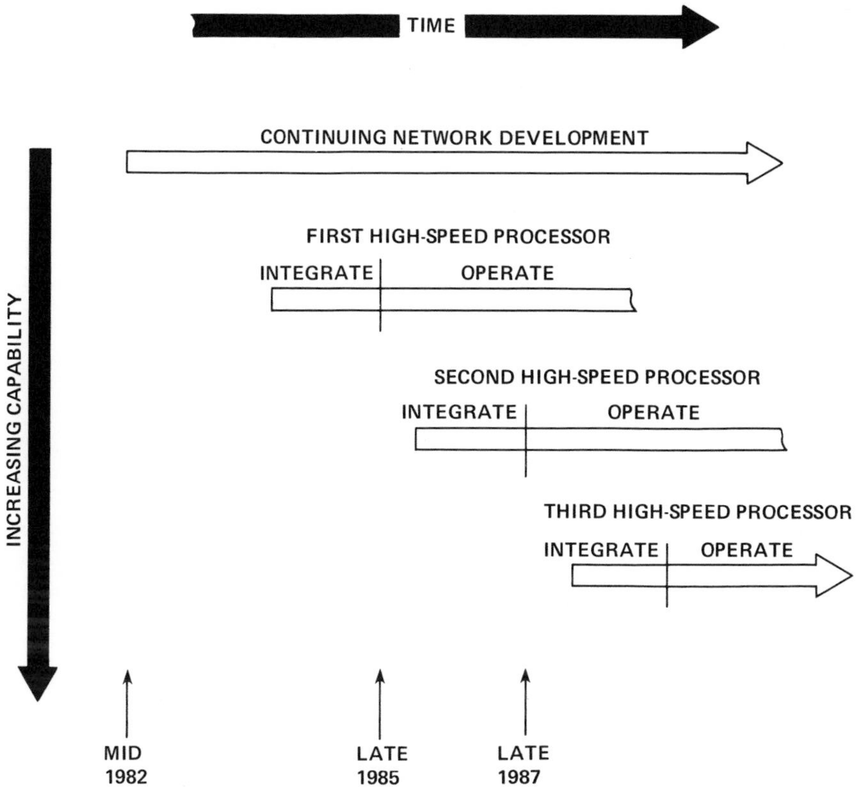

Fig. 3. Evolution of the NAS System.

implemented and tested, followed by an operational phase. An important element in this evolutionary strategy is the early implementation and testing of a fully functional NPSN designed to accommodate new supercomputers from different vendors with a minimum impact on the existing network architecture and on operational use.

As shown in Figure 4, the NPSN will consist of the following eight subsystems:

1. High-Speed Processor Subsystem (HSP)
2. Support Processing Subsystem (SPS)
3. Workstation Subsystem (WKS)
4. Graphics Subsystem (GRS)
5. Mass Storage Subsystem (MSS)
6. Long-Haul Communications Subsystem (LHCS)
7. High-Speed Data-Network Subsystem (HSDN)

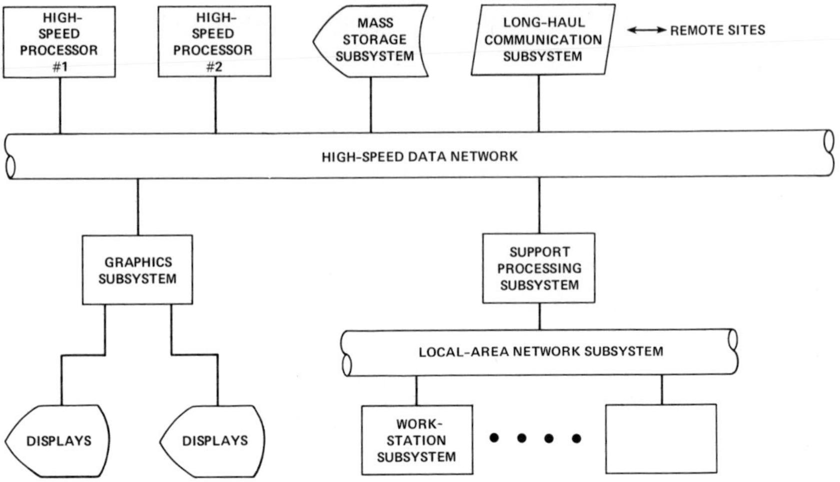

Fig. 4. Schematic of Fully Developed NAS Processing System Network.

8. Local-Area Network Subsystem (LANS)

Only the HSP, SPS, WKS, and GRS will be programmed by users. The MSS and LHCS will provide a mass storage facility and a remote data-communications interface, respectively. The HSDN and LANS will provide a data-transport function for the other subsystems.

4.1 High-Speed Processor Subsystem (HSP)

The HSP is the advanced scientific computing resource within the NPSN. The purpose of this subsystem is to provide the computational throughput and memory capacity to compute computational aerodynamics simulation models. In addition to batch processing, interactive time-sharing processing will be provided to aid in application debugging, result editing, and other activities that depend on close user-processing coupling to achieve optimum overall productivity.

Present plans call for two generations of HSP computers to be in the system at any one time. The first (HSP-1), planned for integration in 1985, will provide a capability to process optimally structured computational aerodynamics applications at a sustained rate of 250 MFLOPS with a minimum 64×10^6 word memory capacity. The second (HSP-2), planned for integration in 1987, will increase these values to 1000 MFLOPS and a 256×10^6 word memory capacity.

4.2 Workstation Subsystem (WKS)

Whereas the HSP is the ultra-high-speed, large-scale computer resource serving the global user community, the WKS is the microprocessor-based resource used by the individual researcher. The WKS will provide a "scientist's work-bench" for local users to perform text and data editing, to process and view graphics files, and to perform small-scale applications processing. Each individual workstation will have the appropriate memory, disk storage, and hard-copy resources to fit the local user's needs. Individual clusters of workstations will be networked together via the LANS for use within local user groups. In addition to local processing, the WKS will provide terminal access to other user-programmed systems and a file-transfer capability via the LANS and HSDN.

4.3 Support Processing Subsystem (SPS)

The SPS is a multi-super-minicomputer-based system providing a number of important functions. The SPS will provide general-purpose interactive processing for local and remote terminal-based users (i.e., those without workstations), and provide an intermediate performance resource between the HSP and WKS performance as a WKS backup. The SPS will be a gateway between the HSDN and the LANS, the location for unit record input/output devices such as high-speed printers and microfilm, and the focal point for network monitoring and system operation.

4.4 Graphics Subsystem (GRS)

The GRS is a super-minicomputer-based system that will provide a sophisticated state-of-the-art graphical-display capability for those applications requiring highly interactive, high-density graphics for input preparation and result analysis. The GRS will provide a level of performance and storage capability beyond that provided by workstations and will be shared by first-level user organizations.

4.5 Mass Storage Subsystem (MSS)

The MSS will provide the global on-line and archival file storage capability for the NPSN. This subsystem will validate and coordinate requests for files to be stored or retrieved within the NPSN and maintain a directory of all contained data.

The MSS will act as a file server for other NPSN subsystems; control its own internal devices; and perform file duplication, media migration, storage allocation, accounting, and file management functions.

Users of the NPSN will create and use files on various subsystems, e.g., HSP, SPS, GRS, or WKS. However, after the user has exited the NPSN, the main repository of these files will be the MSS. This subsystem will hold those very large files that will be used as input to, or generated as output from, the largest tasks that will be processed on the HSP and GRS. It will contain user source and object codes, and parameter and data files that are kept for any significant length of time. The MSS will also contain the backup files that are created to improve the probability that long-lasting or high-value files are accessible when needed.

4.6 Long-Haul Communication Subsystem (LHCS)

The LHCS will provide the data-communication interface between the NPSN and data-communication links to sites geographically remote from Ames Research Center. This subsystem will provide the necessary hardware/software interfaces; modulation and demodulation devices; and recording, processing, data buffering, and management functions to support data transfers and job control by remote users.

In the sense that the MSS is a back-end resource for the entire NPSN, the LHCS is a front-end resource. It provides for access by remote users to the HSP, SPS, GRS, and WKS but it is not specifically addressed or programmed by the user. The LHCS processor functions as a data-communications front-end providing store and forward, protocol conversion, and data-concentrator service.

Current plans call for the LHCS to interface with data links capable of providing 9600 bits/sec to over 1.5×10^6 bits/sec transmission rates in order to interface with government-sponsored networks (e.g., ARPANET, and the proposed NASA Program Support Communication Network) and commercial tariffed services. Candidate data-communication protocols to be supported include ARPANET, IBM System Network Architecture (SNA), and Digital Equipment Corporation's network (DECnet).

4.7 High-Speed Data-Network Subsystem (HSDN)

The HSDN provides the medium over which data and control messages are exchanged among the HSP, SPS, GRS, MSS, and LHCS. Major design emphasis will be placed on the ability to support large file transfers among these systems. The HSDN will include high-speed (minimum 50 megabits/sec) interface devices and driver-level network software to support NPSN internal data communications.

4.8 Local-Area Network Subsystem (LANS)

The LANS will provide the physical data transfer path between the SPS and WKS, and between various workstations within a WKS cluster. The LANS will be designed to support up to 40 workstations and to provide a hardware data-communications rate of at least 10 megabits/sec (e.g., Ethernet).

The LANS differs from the HSDN in data-communication bandwidth because of the smaller size of files transferred on the LANS and the lower cost per LANS network interface device. The HSDN and LANS will use the same network protocol.

5. NPSN SOFTWARE

The NPSN will include a rich set of systems and utility software aimed at providing the most efficient and user-productive environment practical. Major software objectives are as follows:

1. A vendor-independent environment that allows the incorporation of new technologies without sacrificing existing software commitments

2. Common and consistent user environments across the NPSN

3. The maximum transparency to the heterogeneous subsystem nature of the NPSN

4. The optimal performance for critical resources

5. A rich set of user tools

The strategy taken to satisfy these objectives is to use a UNIX™* or UNIX™ look-alike operating system on the user

*The use of trade names of manufacturers in this report does not constitute an official endorsement of such products or manufacturers, either expressed or implied, by the National Aeronautics and Space Administration.

programmable subsystems (HSP, SPS, GRS, and WKS). This approach is the first attempt to achieve vendor independence and a common user environment, and will be implemented by a combination of native UNIX™, vendor UNIX™ look-alikes, and virtual operating system approaches using Software Tools. The UNIX™ approach provides for the implementation of a rich set of user tools from text editors and compilers to graphics packages that are transportable among systems. This approach also provides a degree of transparency among subsystems. Vendor independence and high performance are aided by the implementation of a highly functional and efficient network protocol, such as the Livermore Interactive Network Communication System protocol.

As the NPSN evolution continues, further gains in meeting these objectives in the areas of automatic-file and data-format conversion, common network directory, architecture-independent programming languages, and architecture-specific optimization from ANSI languages will be forthcoming.

6. PROJECTED USAGE ALLOCATION

A NAS system usage policy has been proposed and will be implemented when the initial configuration becomes operational in 1985. It is expected that 90% of the available usage would be allocated to the disciplines of fluid dynamics and aerodynamics. The remaining 10% would support advances in such disciplines as chemistry, atmospheric modeling, aircraft structural analysis, and astrophysics. About 55% of the high-speed processor time would support NASA programs and would be used by the in-house staff and organizations involved in NASA grants, contracts, and joint programs with other organizations. Roughly 20% of the high-speed processor time would support Department of Defense activities, 15% would be allocated to industry, and 5% would be allocated to universities. Costs for industry proprietary work would be reimbursed to the government. The reimbursement formula would include operating and capital improvement costs. This policy is consistent with similar NASA guidelines governing wind-tunnel usage. Usage requests would be screened by Ames line management in a manner similar to the review process governing the Unitary Plan Wind Tunnel operations. Potential users must demonstrate a requirement for the unique capabilities offered by the NAS system. In addition, task selection will be guided by the following criteria: technical quality, national need, timeliness, and funding support.

The guidelines for usage allocation described here apply only to NAS computers in an operational status. During the time that new computers are being integrated into the system, they will be devoted primarily to systems integration, testing, and software development.

7. BENCHMARK TEST SET

Sets of benchmark codes will be constructed to support the system integration and test activity. They will be designed to stress each of the subsystems, which is their primary purpose. However, these codes will also be designed to support pioneering scientific and engineering investigations. This opportunity provides a strong impetus for computational physics researchers to take an active interest in the development of the Benchmark Test Set and in the systems development in general.

Examples of candidate problems under consideration for the Benchmark Test Set include the following: (1) vortex formation, roll-up, and interaction with lifting surfaces; (2) effects of wall motion and geometry on turbulent skin friction; (3) simulation of boundary-layer transition; (4) simulation of afterbody aerodynamics, including effects of real-gas exhaust plumes; (5) simulation of observed nonlinear transonic aeroelastic instabilities of the B-1 wing; (6) numerical optimization of the wing for a transonic transport aircraft; (7) Aero-assisted Orbital Transfer Vehicle (AOTV) flow-field simulation, including real-gas and nonequilibrium radiation physics; (8) complete shuttle orbiter flow-field characteristics; and (9) simulation of crack initiation and propagation. These problems represent a balance of fundamental fluid physics research and applied computational aerodynamics and aerothermodynamics, except the last example listed, which is a computational chemistry investigation.

One of the fluid physics investigations is highlighted in Fig. 5. The basic concept involves the manipulation of the turbulent structure in the vicinity of a wall (or aerodynamic

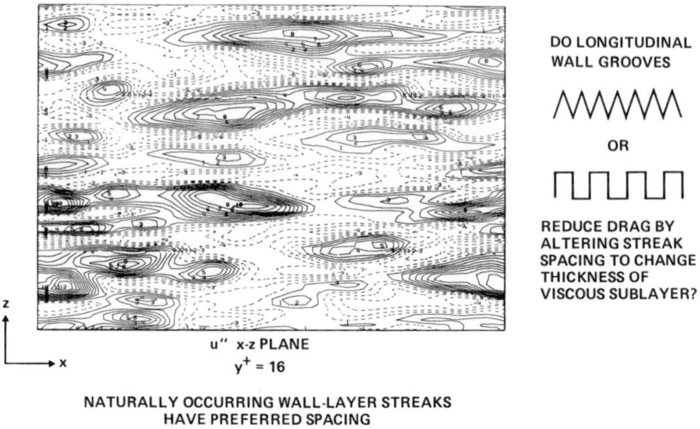
DO LONGITUDINAL
WALL GROOVES
OR
REDUCE DRAG BY
ALTERING STREAK
SPACING TO CHANGE
THICKNESS OF
VISCOUS SUBLAYER?
z
x
u" x-z PLANE
$y^+ = 16$
NATURALLY OCCURRING WALL-LAYER STREAKS
HAVE PREFERRED SPACING

Fig. 5. Candidate Fluid Physics Problem Involving the Manipulation of Turbulence Structure Near a Boundary.

surface) to achieve net skin-friction reduction. Of particular interest is the riblet or wall-groove approach, in which finely spaced longitudinal grooves are inserted on the wall surface to alter the formation and growth of turbulent sublayer streaks. Experimental investigations indicate up to a 20% drag reduction, which would produce enormous benefits in fuel savings and an increased range for transport aircraft. To alter the formation of sublayer streaks, the spacing of the grooves should be of the same order as the individual streak dimensions. Even at moderate Reynolds numbers, a large number of grid points are needed to resolve the riblets and the flow about them, requiring an order-of-magnitude increase in computer power over the Class VI machines. The code would be virtually entirely vectorizable. Storage requirements are 23 words/grid point for a total of 120×10^6 words of data that must be transferred to central memory at each integration time step. A three-dimensional, large-memory color-graphics capability is also needed.

One of the candidate problem examples is illustrated in Fig. 2. The object in this work is to compute the flow field about boattailed afterbodies containing propulsive jets and to study the interactive influence of highly underexpanded exhaust plumes on afterbody drag. The aerodynamic drag associated with jet-engine afterbodies represents a significant fraction of the recoverable drag inherent in the flight of modern aircraft. Because of the very strong interactive and complex nature of the flow, in-depth analysis has not been possible with current computer power. Expensive experimental investigations have been conducted to provide an understanding of the complex flow patterns involved. Even these studies, however, are not sufficiently detailed to adequately describe the strong interaction between the supersonic propulsive jet and the external flow at the base of the afterbody. This interaction produces flow regions of very rapid expansion accompanied by complex recompression patterns and contact discontinuities inside the exhaust-plume outer boundary. The flow over the afterbody separates in a three-dimensional pattern, contributing to the drag. Powerful computational resources would be required to simulate this flow field at flight Reynolds numbers. Typically, discretization of the computational control volume would require about 1×10^6 grid points and 17×10^6 words of memory. The coding is virtually entirely vectorizable, with vector lengths of about 55,000.

These two examples are representative of the candidate list for the NAS Benchmark Test Set. The final test set will be identified at a later time, after extensive analysis has been made to determine the suitability of each of the candidate codes for testing the various NAS subsystems.

8. CONCLUDING REMARKS

Computers are beginning to have a profound effect on aerodynamics, a technical discipline important to the design of all aerospace vehicles. Computational aerodynamics, although severely limited by available computer power, is already emerging as an important aircraft design tool. Successively more refined aerodynamic simulations are resulting from increases in computer power. Strong motivating factors to accelerate the development of computational aerodynamics exist, but a computational system having performance far in excess of that available today is required. The goal of NASA's NAS program is to provide an advanced capability by the mid to late 1980s that will be available for use by government laboratories, industry, and academia. This capability will be continually upgraded beyond that date as computer technology advances.

The currently planned NAS processing system network is based on about 8 yr of study. Each of eight elements of the system has been defined by a set of initial performance goals. In addition, network software strategies for achieving vendor independence and common user environments have been developed. A NAS system usage policy has been proposed for implementation when the initial configuration becomes operational before 1986. Finally, candidate problems in several disciplines besides aerodynamics have been identified for the benchmark test codes.

REFERENCES

1. Magnus, R. and Yoshihara, H., Inviscid transonic flow over airfoils, AIAA J. 8 (1970), 2157-2162.
2. Boppe, C. W. and Stern, M. A., Simulated transonic flows for aircraft with nacelles, pylons and winglets, AIAA Paper 80-0130, Jan. 1980.
3. MacCormack, R. W., Numerical solutions of the interaction of a shock wave with a laminar boundary layer, in Lecture Notes in Physics, vol. 8, Springer Verlag, 1971, 151-163.
4. Deiwert, G. S., A computational investigation of supersonic axisymmetric flow over boattails containing a centered propulsive jet, AIAA Paper 83-0462, Jan. 1983.
5. Agrell, J. and White, R. A., An experimental investigation of supersonic axisymmetric flow over boattails containing a centered propulsive jet, FFA Tech. Note AU-913, 1974.

NASA Ames Research Center
Moffett Field, California 94035

ALGEBRAIC MESH GENERATION FOR LARGE SCALE VISCOUS-COMPRESSIBLE AERODYNAMIC SIMULATION

Robert E. Smith

1. INTRODUCTION.

Viscous-compressible aerodynamic simulation is the numerical solution of the compressible Navier-Stokes equations and associated boundary conditions. Boundary-fitted coordinate systems are well suited for the application of finite difference techniques to the Navier-Stokes equations. An algebraic approach to boundary-fitted coordinate systems is one where an explicit functional relation describes a mesh on which a solution is obtained. This approach has the advantage of rapid-precise mesh control. The basic mathematical structure of three algebraic mesh generation techniques is described. They are transfinite interpolation, the multi-surface method, and the two-boundary technique. The Navier-Stokes equations are transformed to a computational coordinate system where boundary-fitted coordinates can be applied. Large-scale computation implies that there is a large number of mesh points in the coordinate system. Computation of viscous compressible flow using boundary-fitted coordinate systems and the application of this computational philosophy on a vector computer are presented.

The numerical solution of fluid flow about complex geometries using finite difference techniques is dependent on the coordinate systems in which the equations of motion are cast. A finite difference technique is applied on a mesh which is an ordered set of discrete points in a coordinate

ISBN 0-12-546080-5

system. Therefore, solution accuracy and computational expediency depend on the mesh as well as the finite difference technique. Accurate representation of boundary conditions and the resolution of the state variables in regions of rapid change such as boundary layers, shocks, and separated flows, require particular characteristics of the mesh. First, the mesh should adapt to the physical boundaries so that boundary conditions can be applied without special consideration such as interpolation. Secondly, the mesh should be concentrated in regions of rapid change to accurately capture the solution there. From a computational point of view, it is desirable that the mesh be uniform, the boundaries enclose a rectangular parallelepiped (Fig. 1), and the exterior boundaries of the

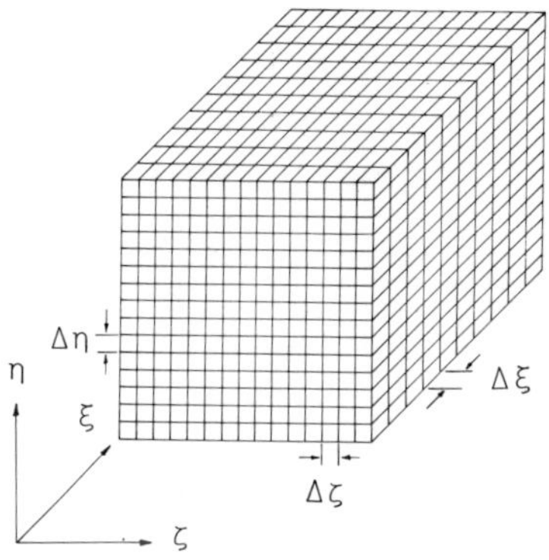

Fig. 1. Computational mesh.

parallelepiped correspond to physical boundaries. If this is possible, overall computer program logic can be minimized, and the application of the finite difference technique can be kept highly repetitive over the entire mesh. Repetitious computation is particularly important when a computer with vector architecture [1, 2] is used to obtain a solution.

For a large class of problems, an approach which tends to satisfy both the accuracy and expediency requirements is to transform the equations of motion from the originally defined Cartesian coordinates, referred to as the physical coordinates, to uniform rectangular coordinates, referred to as the computational coordinates. The particular region of a coordinate system where a mesh is defined is referred to as a domain.

A transformation between the physical domain and the computational domain (Fig. 2) is a mathematical relationship mapping one domain into the other. Similarly the mesh in one domain is mapped into the mesh in the other domain. When the transformation maps boundaries in the physical domain into boundaries in the computational domain the term "boundary-fitted coordinate system" is used to describe the transformation.

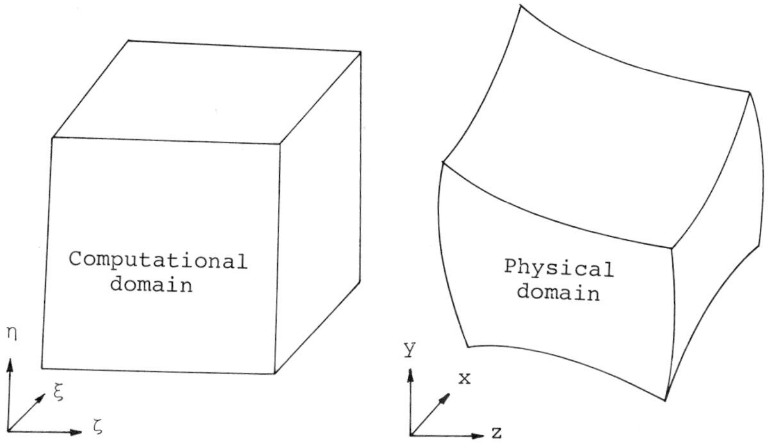

Fig. 2. Computational domain--physical domain.

An indirect (differential) approach for finding the relationship between the computational and physical meshes, due to Thompson et al. [3, 4, 5] has been highly successful. In this approach the elliptic system of partial differential equations [3] which the relationship between the two domains must satisfy is numerically solved by an iterative technique such as Successive-Over-Relaxation (SOR). The numerical solution is the mesh in the physical domain corresponding to the mesh in the computational domain. A direct (algebraic) approach, where an explicit functional relationship between the computational domain and the physical domain is known has the advantage that changes to the mesh are direct and rapidly obtained. Also, a resolved mesh with few mesh points is valid when expanded to a large number of points. As stated previously, this paper describes three algebraic mesh generation techniques. They are transfinite interpolation, the multisurface method, and the two-boundary technique. Portions of the description follow the author presentation in reference 6. This is preceded by a description of boundary-fitted coordinate transformations and the application of these transformations in the Navier-Stokes equations. The application of boundary-fitted coordinates to compute viscous compressible flow on a vector computer follows the description of the algebraic methods.

2. BOUNDARY-FITTED COORDINATE TRANSFORMATIONS.

A transformation between a computational domain and a physical domain is a unique single-valued functional relation. This is represented symbolically by letting x, y, and z be coordinates in the physical domain and ξ, η, and ζ be coordinates in the computational domain, then

$$\xi = \xi(x,y,z), \quad \eta = \eta(z,y,z), \quad \zeta = \zeta(x,y,z),$$

and conversely

$$x = x(\xi,\eta,\zeta), \quad y = y(\xi,\eta,\zeta), \quad z = z(\xi,\eta,\zeta).$$

The bounds of the computational domain are defined by

$$0 \le \xi \le 1,$$

$$0 \le \eta \le 1,$$

$$0 \le \zeta \le 1.$$

A uniform mesh (Fig. 1) is superimposed onto the computational domain by letting

$$\Delta\xi = \text{constant}_1,$$

$$\Delta\eta = \text{constant}_2,$$

$$\Delta\zeta = \text{constant}_3.$$

Given the functional relations

$$x = x(\xi,\eta,\zeta), \quad y = y(\xi,\eta,\zeta), \text{ and } \quad z = z(\xi,\eta,\zeta)$$

the uniform mesh in the computational domain is transformed to a corresponding mesh in the physical domain.

In order to transform the equations of motion, partial derivatives with respect to the independent variables x, y, and z must be transformed to partial derivatives with respect to the variables ξ, η, and ζ. For example, if

u, v, and w are velocities in the x, y, and z directions in the equations of motion, then the first derivatives of u, v, and w with respect to x, y, and z are transformed to first derivatives with respect to ξ, η, and ζ by chain differentiation. That is

$$\begin{bmatrix} \frac{\partial u}{\partial x} & \frac{\partial u}{\partial y} & \frac{\partial u}{\partial z} \\ \frac{\partial v}{\partial x} & \frac{\partial v}{\partial y} & \frac{\partial v}{\partial z} \\ \frac{\partial w}{\partial x} & \frac{\partial w}{\partial y} & \frac{\partial w}{\partial z} \end{bmatrix} = \begin{bmatrix} \frac{\partial u}{\partial \xi} & \frac{\partial u}{\partial \eta} & \frac{\partial u}{\partial \zeta} \\ \frac{\partial v}{\partial \xi} & \frac{\partial v}{\partial \eta} & \frac{\partial v}{\partial \zeta} \\ \frac{\partial w}{\partial \xi} & \frac{\partial w}{\partial \eta} & \frac{\partial w}{\partial \zeta} \end{bmatrix} \begin{bmatrix} \frac{\partial \xi}{\partial x} & \frac{\partial \xi}{\partial y} & \frac{\partial \xi}{\partial z} \\ \frac{\partial \eta}{\partial x} & \frac{\partial \eta}{\partial y} & \frac{\partial \eta}{\partial z} \\ \frac{\partial \zeta}{\partial x} & \frac{\partial \zeta}{\partial y} & \frac{\partial \zeta}{\partial z} \end{bmatrix}. \quad (2.1)$$

The matrix

$$J = \begin{bmatrix} \frac{\partial \xi}{\partial x} & \frac{\partial \xi}{\partial y} & \frac{\partial \xi}{\partial z} \\ \frac{\partial \eta}{\partial x} & \frac{\partial \eta}{\partial y} & \frac{\partial \eta}{\partial z} \\ \frac{\partial \zeta}{\partial x} & \frac{\partial \zeta}{\partial y} & \frac{\partial \zeta}{\partial z} \end{bmatrix} = \begin{bmatrix} \xi_x & \xi_y & \xi_z \\ \eta_x & \eta_y & \eta_z \\ \zeta_x & \zeta_y & \zeta_z \end{bmatrix}$$

is the Jacobian matrix of the transformation. If the functional relations

$$\xi = \xi(x,y,z), \quad \eta = \eta(x,y,z), \text{ and } \quad \zeta = \zeta(x,y,z)$$

are known, then the Jacobian matrix can be directly found by differentiating the functions. It is not necessary, however, to explicitly know ξ, η, and ζ as functions of x, y, and z to determine the Jacobian matrix. The inverse Jacobian matrix

$$J^{-1} \equiv \begin{bmatrix} \frac{\partial x}{\partial \xi} & \frac{\partial x}{\partial \eta} & \frac{\partial x}{\partial \zeta} \\ \frac{\partial y}{\partial \xi} & \frac{\partial y}{\partial \eta} & \frac{\partial y}{\partial \zeta} \\ \frac{\partial z}{\partial \xi} & \frac{\partial z}{\partial \eta} & \frac{\partial z}{\partial \zeta} \end{bmatrix}$$

can be obtained by differentiating the functions

$$x = x(\xi,\eta,\zeta), \quad y = y(\xi,\eta,\zeta), \text{ and } \quad z = z(\xi,\eta,\zeta)$$

with respect to ξ, η, and ζ. With these derivatives

$$J = \frac{\text{Transposed of Cofactor } (J^{-1})}{|J^{-1}|},$$

where $|J^{-1}|$ is the Jacobian determinate.

$$|J^{-1}| = \begin{vmatrix} \frac{\partial x}{\partial \xi} & \frac{\partial x}{\partial \eta} & \frac{\partial x}{\partial \zeta} \\ \frac{\partial y}{\partial \xi} & \frac{\partial y}{\partial \eta} & \frac{\partial y}{\partial \zeta} \\ \frac{\partial z}{\partial \xi} & \frac{\partial z}{\partial \eta} & \frac{\partial z}{\partial \zeta} \end{vmatrix}$$

$$= \frac{\partial x}{\partial \xi}\left(\frac{\partial y}{\partial \eta}\frac{\partial z}{\partial \zeta} - \frac{\partial y}{\partial \zeta}\frac{\partial z}{\partial \eta}\right) - \frac{\partial x}{\partial \eta}\left(\frac{\partial y}{\partial \xi}\frac{\partial z}{\partial \zeta} - \frac{\partial y}{\partial \zeta}\frac{\partial z}{\partial \xi}\right)$$

$$+ \frac{\partial x}{\partial \zeta}\left(\frac{\partial y}{\partial \xi}\frac{\partial z}{\partial \eta} - \frac{\partial y}{\partial \eta}\frac{\partial z}{\partial \xi}\right) \tag{2.2}$$

Thus

$$J = \frac{1}{|J^{-1}|}\begin{bmatrix} \left(\frac{\partial y}{\partial \eta}\frac{\partial z}{\partial \zeta} - \frac{\partial y}{\partial \zeta}\frac{\partial z}{\partial \eta}\right) & -\left(\frac{\partial x}{\partial \eta}\frac{\partial z}{\partial \zeta} - \frac{\partial x}{\partial \zeta}\frac{\partial z}{\partial \eta}\right) & \left(\frac{\partial x}{\partial \eta}\frac{\partial y}{\partial \zeta} - \frac{\partial x}{\partial \zeta}\frac{\partial y}{\partial \eta}\right) \\ -\left(\frac{\partial y}{\partial \xi}\frac{\partial z}{\partial \zeta} - \frac{\partial y}{\partial \zeta}\frac{\partial z}{\partial \xi}\right) & \left(\frac{\partial x}{\partial \xi}\frac{\partial z}{\partial \zeta} - \frac{\partial x}{\partial \zeta}\frac{\partial z}{\partial \xi}\right) & -\left(\frac{\partial x}{\partial \xi}\frac{\partial y}{\partial \zeta} - \frac{\partial x}{\partial \zeta}\frac{\partial y}{\partial \xi}\right) \\ \left(\frac{\partial y}{\partial \xi}\frac{\partial z}{\partial \eta} - \frac{\partial y}{\partial \eta}\frac{\partial z}{\partial \xi}\right) & -\left(\frac{\partial x}{\partial \xi}\frac{\partial z}{\partial \eta} - \frac{\partial x}{\partial \eta}\frac{\partial z}{\partial \xi}\right) & \left(\frac{\partial x}{\partial \xi}\frac{\partial y}{\partial \eta} - \frac{\partial x}{\partial \eta}\frac{\partial y}{\partial \xi}\right) \end{bmatrix} \tag{2.3}$$

provided $|J^{-1}| \neq 0$. Higher derivative analysis can be pursued in a similar fashion. For more information on the transformation of partial differential equations the reader is referred to References 1 and 2.

It is rare to find algebraic expressions of the computational coordinates as functions of the physical coordinates. The preferred approach is to express the physical domain as a function of the computational domain and differentiate the physical mesh with respect to the computational mesh. It is very important that derivative evaluation be performed and incorporated into the finite difference approximation of the equations of motion in such a manner that geometrically-induced errors are not created. References 7 and 8 address this subject.

3. NAVIER-STOKES EQUATIONS.

The unsteady compressible Navier-Stokes equations in three dimensions can be written in conservation form [1]

$$U_{\hat{t}} + F_x + G_y + H_z = 0 \tag{3.1}$$

where

$$U = \begin{pmatrix} \rho \\ \rho u \\ \rho v \\ \rho w \\ \rho e \end{pmatrix}, \quad F = \begin{pmatrix} \rho \\ \rho u^2 - \tau_{xx} \\ \rho uv - \tau_{xy} \\ \rho uw - \tau_{xz} \\ \rho ue + \dot{q}_x - \phi_x \end{pmatrix},$$

$$G = \begin{pmatrix} \rho v \\ \rho uv - \tau_{xy} \\ \rho v^2 - \tau_{yy} \\ \rho vw - \tau_{yz} \\ \rho ve + \dot{q}_y - \phi_y \end{pmatrix}, \quad H = \begin{pmatrix} \rho w \\ \rho wu - \tau_{xz} \\ \rho wv - \tau_{yz} \\ \rho w^2 - \tau_{zz} \\ \rho we + \dot{q}_z - \phi_z \end{pmatrix},$$

$$\begin{pmatrix} \tau_{xx} & \tau_{xy} & \tau_{xz} \\ \tau_{xy} & \tau_{yy} & \tau_{yz} \\ \tau_{xz} & \tau_{yz} & \tau_{zz} \end{pmatrix} \equiv \text{Stress tensor}, \quad \begin{pmatrix} \dot{q}_x \\ \dot{q}_y \\ \dot{q}_z \end{pmatrix} = \hat{k} \begin{pmatrix} T_x \\ T_y \\ T_z \end{pmatrix},$$

$\hat{k}$ = Coefficient of thermal conductivity,

$$\begin{pmatrix} \phi_x \\ \phi_y \\ \phi_z \end{pmatrix}^T = (u \quad v \quad w) \begin{pmatrix} \tau_{xx} & \tau_{xy} & \tau_{xz} \\ \tau_{xy} & \tau_{yy} & \tau_{yz} \\ \tau_{xz} & \tau_{yz} & \tau_{zz} \end{pmatrix}.$$

There are five equations with seven unknowns. The unknowns are $\rho \equiv$ Density, $u,v,w \equiv$ Velocity components; $e \equiv$ Energy ($T \equiv$ Temperature); $\mu \equiv$ Viscosity coefficient; and $P \equiv$ Pressure. An equation of state ($P = \rho RT$) and Sutherland viscosity law provide the two additional equations to match the number of equations with the number of unknowns.

The governing equations are written relative to a Cartesian coordinate system where the physical domain is defined. A transformation of the governing equations is accomplished by defining the computational domain by

$$0 \leq \xi \leq 1,$$

$$0 \leq \eta \leq 1,$$

$$0 \leq \zeta \leq 1,$$

and expressing

$$\begin{bmatrix} u_x & u_y & u_z \\ v_x & v_y & v_z \\ w_x & w_y & w_z \\ \rho_x & \rho_y & \rho_z \\ T_x & T_y & T_z \\ P_x & P_y & P_z \end{bmatrix} = \begin{bmatrix} u_\xi & u_\eta & u_\zeta \\ v_\xi & v_\eta & v_\zeta \\ w_\xi & w_\eta & w_\zeta \\ \rho_\xi & \rho_\eta & \rho_\zeta \\ T_\xi & T_\eta & T_\zeta \\ P_\xi & P_\eta & P_\zeta \end{bmatrix} \begin{bmatrix} \xi_x & \xi_y & \xi_z \\ \eta_x & \eta_y & \eta_z \\ \zeta_x & \zeta_y & \zeta_z \end{bmatrix}.$$

The governing equations become

$$U_t + (F_\xi F_\eta F_\zeta)\begin{pmatrix} \xi_x \\ \xi_y \\ \xi_z \end{pmatrix} + (G_\xi G_\eta G_\zeta)\begin{pmatrix} \eta_x \\ \eta_y \\ \eta_z \end{pmatrix} + (H_\xi H_\eta H_\zeta)\begin{pmatrix} \zeta_x \\ \zeta_y \\ \zeta_z \end{pmatrix}$$
$$= 0. \tag{3.2}$$

A typical term

$$\tau_{xx} = -P + 2\mu u_x - \frac{2}{3}\mu\,[u_x + u_y + u_z]$$

becomes

$$\tau_{xx} = -P + 2\mu(\mu_\xi \xi_x + u_\eta \eta_x + u_\zeta \zeta_x) - \frac{2}{3}\mu(\mu_\xi \xi_x + u_\eta \eta_x$$
$$+ u_\zeta \zeta_x + v_\xi \xi_x + v_\eta \eta_x + v_\zeta \zeta_x + w_\xi \xi_x + w_\eta \eta_x + w_\zeta \zeta_z).$$

The matrix $\begin{pmatrix} \xi_x \xi_y \xi_z \\ \eta_x \eta_y \eta_z \\ \zeta_x \zeta_y \zeta_z \end{pmatrix} \equiv J$ is the Jacobian matrix and must be determined prior to the solution of the governing equations.

4. ALGEBRAIC TECHNIQUES FOR MESH GENERATION.

Transfinite interpolation [9] described by Gordon and Hall in the early 1970's is a highly generalized algebraic mesh generation method. Transfinite interpolation is applied through a series of univariate interpolations where blending functions and the associated parameters (point position and/or derivatives) determine a mesh. For aerodynamic application Eriksson [10] and Rizzi and Eriksson [11] have adopted the original transfinite interpolation formulation to use only exterior boundary descriptions and derivatives at certain boundaries. They have also incorporated exponentials into the blending functions to concentrate the mesh near an exterior boundary. The multisurface method [12, 13] developed by Peter Eiseman provides formulas for mesh definition based on mesh descriptions of two boundary surfaces and an arbitrary number of intermediate control surfaces. Choosing interpolants (defined similar to blending functions) and the placement of the control surfaces determines mesh shape and spacing. The multisurface method has been used by Eiseman in numerous applications [14, 15] but most notably for computing meshes about turbine cascades. The two-boundary technique [1, 2, 16] described by this author is based on the description of two exterior boundaries and the application of either linear or hermite cubic polynomial interpolation to compute the interior mesh. For cubic interpolation, surface derivatives combined with magnitude coefficients control the orthogonality of the mesh at and near the boundaries.

4.1 Transfinite interpolation.

A transformation from the computational domain to the physical domain is a vector-valued function

$$\vec{F}(\xi,\eta,\zeta) = \begin{bmatrix} x(\xi,\eta,\zeta) \\ y(\xi,\eta,\zeta) \\ z(\xi,\eta,\zeta) \end{bmatrix} \tag{4.1}$$

where

$$0 \leq \xi \leq 1,$$

$$0 \leq \eta \leq 1,$$

$$0 \leq \zeta \leq 1.$$

Using only the outer boundary surfaces (Fig. 3) and out of surface derivatives at certain boundaries to define an interior mesh [6] is reasonable since normally a great deal of geometric information is known at bounding surfaces, but not always away from them.

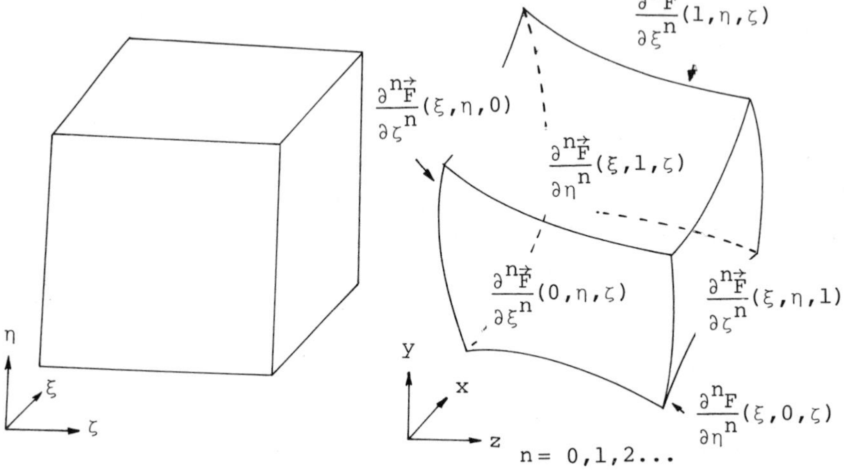

Fig. 3. Transfinite interpolation--outer surface description.

Let

$$\frac{\partial^n \vec{F}}{\partial \xi^n}(\xi_\ell,\eta,\zeta) = \vec{A}_\ell^n(\eta,\zeta), \qquad \ell=1,2 \quad n=0,1\ldots P,$$

$$\frac{\partial^n \vec{F}}{\partial \eta^n}(\varepsilon,\eta_\ell,\zeta) = \vec{B}_\ell^n(\xi,\zeta), \qquad \ell=1,2 \quad n=0,1\ldots Q,$$

$$\frac{\partial^n \vec{F}}{\partial \zeta^n}(\xi,\eta,\zeta_\ell) = \vec{C}_\ell^n(\xi,\eta), \qquad \ell=1,2 \quad n=0,1\ldots R\,.$$

A set of blending functions is defined by

$$\alpha_\ell^{(n)}(\xi), \qquad \ell=1,2, \quad n=0,1\ldots P,$$

$$\beta_\ell^{(n)}(\eta), \qquad \ell=1,2, \quad n=0,1\ldots Q,$$

$$\gamma_\ell^{(n)}(\zeta), \qquad \ell=1,2, \quad n=0,1\ldots R,$$

with the conditions

$$\frac{\partial^m \alpha_\ell^{(n)}(\xi)}{\partial \xi^m} = \delta_{\ell 1}\ \delta_{nm},$$

$$\frac{\partial^m \beta_\ell^{(n)}(\eta)}{\partial \eta^m} = \delta_{\ell 1}\ \delta_{nm},$$

$$\frac{\partial^m \gamma_\ell^{(n)}(\zeta)}{\partial \zeta^m} = \delta_{\ell 1}\ \delta_{nm},$$

and where the δ functions are defined by

$$\delta_{ij} = 0, \quad i \neq j, \quad \delta_{ij} = 1, \quad (i = j).$$

The transfinite interpolation algorithm becomes

$$\vec{F}_1(\xi,\eta,\zeta) = \sum_{\ell=1}^{2} \sum_{n=0}^{P} \alpha_\ell^{(n)}\ \vec{A}_\ell^n(\eta,\zeta), \tag{4.2}$$

$$\vec{F}_2(\xi,\eta,\zeta) = \vec{F}_1(\xi,\eta,\zeta) + \sum_{\ell=1}^{2} \sum_{n=0}^{Q} \beta_\ell^{(n)}(\eta) \left[\vec{B}_\ell^n(\xi,\zeta) - \frac{\partial^n \vec{F}_1}{\partial \eta^n}(\varepsilon,\eta_\ell,\zeta) \right], \tag{4.3}$$

$$F(\xi,\eta,\zeta) = \vec{F}_2(\xi,\eta,\zeta) + \sum_{\ell=1}^{2} \sum_{n=0}^{R} \gamma_\ell^{(n)}(\zeta) \left[\vec{C}_\ell^n(\xi,\eta) - \frac{\partial^n \vec{F}_2}{\partial \zeta^n}(\xi,\eta,\zeta_\ell) \right]. \tag{4.4}$$

The boundary sets are

$$\left\{ x_{\ell JK}, y_{\ell JK}, z_{\ell JK} \right\}_{\substack{K=1\\J=1\\\ell=1}}^{\substack{K=M\\J=N\\\ell=2}}, \quad \left\{ 0, \eta_{JK}, \zeta_{JK} \right\}_{\substack{K=1\\J=1}}^{\substack{K=M\\J=N}}, \quad \left\{ 1, \eta_{JK}, \zeta_{JK} \right\}_{\substack{K=1\\J=1}}^{\substack{K=M\\J=N}},$$

$$\left\{x_{I\ell K}, y_{I\ell K}, z_{I\ell K}\right\}_{\substack{K=1\\ \ell=1\\ I=1}}^{\substack{K=M\\ \ell=2\\ I=L}}, \quad \left\{\xi_{IK}, 0, \zeta_{IK}\right\}_{\substack{K=1\\ I=1}}^{\substack{K=M\\ I=L}}, \quad \left\{\xi_{IK}, 1, \zeta_{IK}\right\}_{\substack{K=1\\ I=1}}^{\substack{K=M\\ I=L}},$$

$$\left\{x_{IJ\ell}, y_{IJ\ell}, z_{IJ\ell}\right\}_{\substack{\ell=1\\ J=1\\ I=1}}^{\substack{\ell=2\\ J=N\\ I=L}}, \quad \left\{\xi_{IJ}, \eta_{IJ}, 0\right\}_{\substack{J=L\\ I=L}}^{\substack{J=N\\ I=L}}, \quad \left\{\xi_{IJ}, \eta_{IJ}, 1\right\}_{\substack{J=1\\ I=1}}^{\substack{J=N\\ I=L}}.$$

Also, outward derivatives at certain of these boundaries as well as the blending functions are required for this formulation. Deriving appropriate blending functions is the key element and it can vary from one mesh problem to another.

4.2 The multisurface method

The multisurface method [12, 13] is a procedure for generating coordinates between an inner boundary surface $\vec{S}_1(\xi,\zeta)$ and an outer boundary surface $\vec{S}_N(\xi,\zeta)$. An arbitrary number of internal surfaces $\vec{S}_2(\xi,\zeta)\ldots\vec{S}_{N-1}(\xi,\zeta)$ are introduced to control the coordinate representation between $\vec{S}_1(\xi,\zeta)$ and $\vec{S}_N(\xi,\zeta)$ (Fig. 4). Each surface representation is such that

$$\vec{S}_k(\xi,\zeta) = \begin{bmatrix} x_k(\xi,\zeta) \\ y_k(\xi,\zeta) \\ z_k(\xi,\zeta) \end{bmatrix}, \quad k=1\ldots N.$$

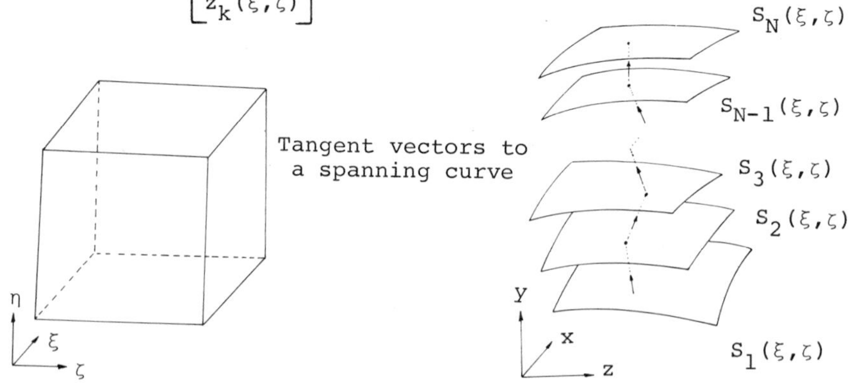

Fig. 4. The multisurface method.

The physical domain can be written as

$$\vec{F}(\xi,\eta,\zeta) = \begin{bmatrix} X(\vec{S}_1(\xi,\zeta),\ \vec{S}_2(\xi,\zeta)\ldots\vec{S}_N(\xi,\zeta),\eta) \\ Y(\vec{S}_1(\xi,\zeta),\ \vec{S}_2(\xi,\zeta)\ldots\vec{S}_N(\xi,\zeta),\eta) \\ Z(\vec{S}_1(\xi,\zeta),\ \vec{S}_2(\xi,\zeta)\ldots\vec{S}_N(\xi,\zeta),\eta) \end{bmatrix} \tag{4.5}$$

where

$$0 \le \xi \le 1, \qquad 0 \le \eta \le 1, \qquad 0 \le \zeta \le 1.$$

The variable η is the independent variable spanning between surfaces.

It is assumed that the set of surfaces described above are ordered from bounding surface to bounding surface, and for a fixed ξ and ζ there is a corresponding point on each surface. The intermediate surfaces are not coordinate surfaces, but instead are surfaces which are used to establish a field of tangent vectors to the coordinate curve spanning across the surfaces. For the time being, it is assumed that the bounding surfaces are coordinate surfaces. A smooth interpolation connecting the bounding surfaces results in a smooth vector field of tangent directions but with unspecified magnitudes. A unique vector field of tangents is obtained by correctly choosing magnitudes which on integration fit precisely the bounding surfaces. This is demonstrated with the vector field of tangents given by

$$\vec{V}_k(\xi,\zeta) = E_k\left[\vec{S}_{k+1}(\xi,\zeta) - \vec{S}_k(\xi,\zeta)\right] \qquad k=1\ldots N,\ (\text{Fig. } 5). \tag{4.6}$$

The coefficients E_k are scalars which determine the magnitude of the vectors but not the direction. Using the independent variable η for the spanning direction, a partition $\eta_1 < \eta_2 \cdots < \eta_{N-1}$ can be specified in correspondence with the tangents in Eq. (4.6). The partitioned variable can be used to represent the tangents as discrete vector-valued functions which map η_k into $\vec{V}_k(\xi,\zeta)$. The first derivative of $\vec{F}(\xi,\eta,\zeta)$ with respect to η is given by

$$\frac{\partial \vec{F}}{\partial \eta}(\xi,\eta,\zeta) = \sum_{k=1}^{N-1} \psi_k(\eta)\ \vec{V}_k(\xi,\zeta) \tag{4.7}$$

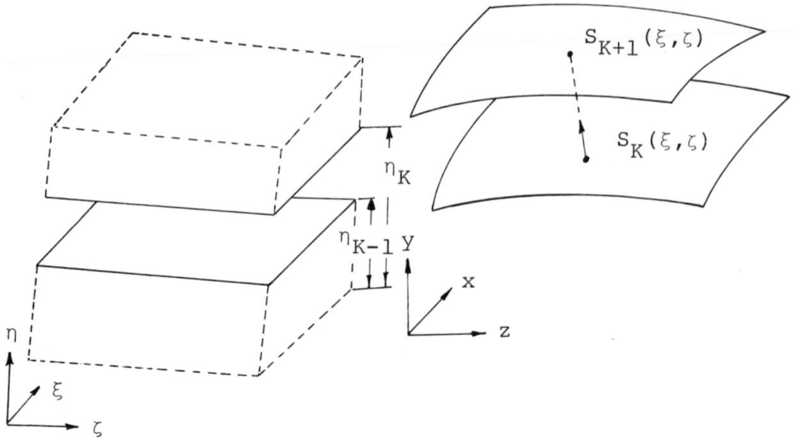

Fig. 5. Tangents to a piecewise linear curve and a partition of the spanning variable from the computational domain.

where

$$\psi_k(\eta_\ell) = \delta_{k\ell},$$

and

$$\delta_{k\ell} = 0 \quad k \neq \ell,$$

$$\delta_{k\ell} = 1 \quad k = \ell.$$

The interpolants $\psi_k(\eta)$ are defined similar to blending functions in Eq. (4.2) but here they are used to describe a derivative function and multiply a tangent vector field. Integrating Eq. (4.7) with an initial η and $\vec{S}_1(\xi,\zeta)$ yields

$$\vec{F}(\xi,\eta,\zeta) = \vec{S}_1(\xi,\zeta) + \sum_{k=1}^{N-1} E_k G_k(\eta)\left[\vec{S}_{k+1}(\xi,\zeta) - \vec{S}_k(\xi,\zeta)\right] \tag{4.8}$$

where

$$G_k(\eta) = \int_{\eta_1}^{\eta} \psi_k(x)\, dx.$$

If the magnitudes E_k are chosen so that each $E_k G_k(\eta_{N-1}) = 1$, then the evaluation of Eq. (4.8) at η_{N-1} reduces to $S_N(\xi,\zeta)$. This allows Eq. (4.8) to be expressed as

$$\vec{F}(\xi,\eta,\zeta) = \vec{S}_1(\xi,\zeta) + \sum_{k=1}^{N-1} \frac{G_k(\eta)}{G_k(\eta_{N-1})} \left[\vec{S}_{k+1}(\xi,\zeta) - \vec{S}_k(\xi,\zeta)\right] \qquad (4.9)$$

which is referred to as Eiseman's general multisurface transformation.

The basic ingredients of the multisurface method are the partition $\eta_1 < \eta_2 \ldots < \eta_{N-1}$, the interpolents ψ_k and the surfaces $\vec{S}_k(\xi,\zeta)$. Choosing ψ_k to be polynomials of degree N in η, the curve connecting the bounding surfaces is of degree $N + 1$. In a systematic fashion

$$\psi_k(\eta) = \prod_{\substack{i=1 \\ i \neq k}}^{N-1} (\eta - \eta_i).$$

Comparing the multisurface method with transfinite interpolation, the multisurface method requires interpolants $\psi_k(\eta)$, one set of surfaces, and allows interpolation in only one coordinate direction. It is apparent that blending functions can be derived from Eiseman transformation formulas starting with interpolants. It is important to remember that the most difficult aspect of algebraic mesh generation is the determination of functions (blending functions, interpolants, etc.) which control a mesh. The emphasis in the multisurface development is on deriving interpolants which provide satisfactory control.

4.3 The two-boundary technique

The two-boundary technique [1, 2, 16] has common characteristics with transfinite interpolation where position and derivatives on exterior boundaries along with blending functions are used to define the physical domain. For the two-boundary technique blending functions are specified to be linear and cubic polynomials, and control functions are incorporated to further enhance mesh spacing control.

The technique is based on defining two nonintersecting surfaces $\vec{S}_1(\xi,\zeta)$ and $\vec{S}_2(\xi,\zeta)$ (Fig. 6) where

$$\vec{S}_1(\xi,\zeta) = \begin{bmatrix} x_1(\xi,\zeta) \\ y_1(\xi,\zeta) \\ z_1(\xi,\zeta) \end{bmatrix}, \qquad \vec{S}_2(\xi,\zeta) = \begin{bmatrix} x_2(\xi,\zeta) \\ y_2(\xi,\zeta) \\ z_2(\xi,\zeta) \end{bmatrix}$$

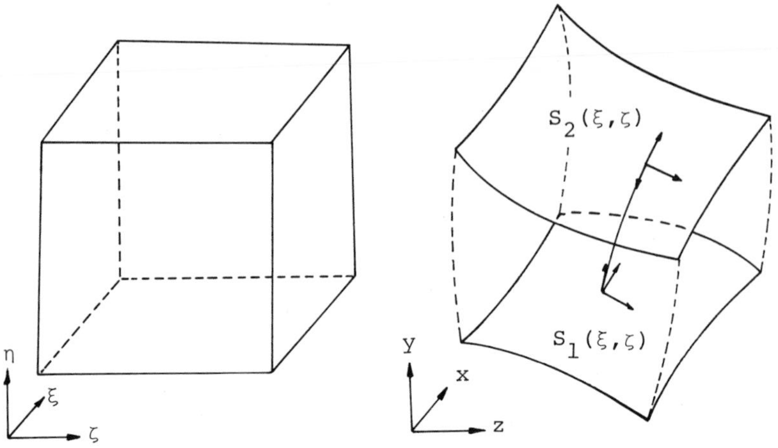

Fig. 6. The two-boundary technique.

and

$$0 \le \xi \le 1, \qquad 0 \le \zeta \le 1.$$

The physical domain is expressed

$$\vec{F}(\xi,\eta,\zeta) = \begin{bmatrix} x(\xi,\eta,\zeta) \\ y(\xi,\eta,\zeta) \\ z(\xi,\eta,\zeta) \end{bmatrix} = \begin{bmatrix} X(\vec{S}_1(\xi,\zeta),\ \vec{S}_2(\xi,\zeta),\eta) \\ Y(\vec{S}_1(\xi,\zeta),\ \vec{S}_2(\xi,\zeta),\eta) \\ Z(\vec{S}_1(\xi,\zeta),\ \vec{S}_2(\xi,\zeta),\eta) \end{bmatrix}$$

Explicit forms of the two-boundary technique are linear and hermite cubic interpolation. The linear form is

$$\vec{F}(\xi,\eta,\zeta) = \sum_{k=1}^{2} \lambda_k(\eta)\vec{S}_k(\xi,\zeta) \tag{4.10}$$

where

$$\lambda_1(\eta) = 1 - \eta,$$

$$\lambda_2(\eta) = \eta.$$

The cubic formulation is

$$\vec{F}(\xi,\eta,\zeta) = \sum_{k=1}^{2} \mu_k(\eta)\vec{S}_k(\xi,\zeta) + \sum_{k=1}^{2} \mu_{k+2}(\eta)\left[\frac{\partial\vec{S}_k}{\partial\xi}(\xi,\zeta) \times \frac{\partial\vec{S}_k}{\partial\zeta}(\xi,\zeta)\right] \tag{4.11}$$

where

$$\mu_1(\eta) = 2\eta^3 - 3\eta^2 + 1,$$

$$\mu_2(\eta) = -2\eta^3 + 3\eta^2,$$

$$\mu_3(\eta) = \eta^3 - 2\eta^2 + 1, \qquad (4.12)$$

$$\mu_4(\eta) = \eta^3 - \eta^2,$$

$$0 \leq \eta \leq 1.$$

The cross product of surface derivatives or normal derivatives at the boundaries is given by

$$\frac{\partial \vec{S}_k}{\partial \xi}(\xi,\zeta) \times \frac{\partial \vec{S}_k}{\partial \zeta}(\xi,\zeta) = T_k \begin{vmatrix} \vec{i} & \vec{j} & \vec{k} \\ \frac{\partial x_k}{\partial \xi}(\varepsilon,\zeta) & \frac{\partial y_k}{\partial \xi}(\varepsilon,\zeta) & \frac{\partial z_k}{\partial \xi}(\varepsilon,\zeta) \\ \frac{\partial x_k}{\partial \zeta}(\varepsilon,\zeta) & \frac{\partial y_k}{\partial \zeta}(\varepsilon,\zeta) & \frac{\partial z_k}{\partial \zeta}(\varepsilon,\zeta) \end{vmatrix}, \quad k=1,2. \qquad (4.13)$$

The constants T_k control the magnitudes of the normal derivatives of the boundaries. For nonzero T_k a mesh resulting from this formulation is orthogonal at the boundaries $\vec{S}_k(\xi,\zeta)$. Increasing the magnitudes T_k forces the effect of orthogonality further into the interior of the physical mesh. If the magnitudes are too large, the mesh becomes double-valued and is unsatisfactory (Fig. 7).

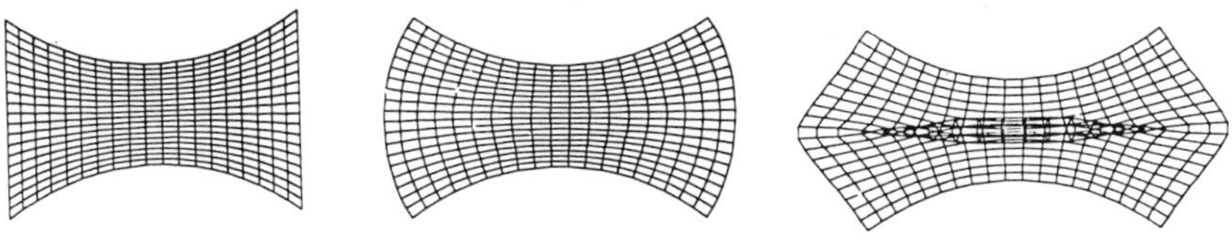

No orthogonality magnitude — Satisfactory orthogonality magnitude — Unsatisfactory orthogonality magnitude

Fig. 7. Mesh orthogonality.

The key ingredients for the two-boundary technique, as it is presented here, are two nonintersecting bounding surfaces, normal magnitude constants or normal magnitude functions. It is later shown that additional ingredients are control functions which govern the spacing of a mesh.

5. MESH SPACING CONTROL.

The spacing of a mesh in the physical domain is primarily affected by how the computational coordinates are incorporated into the blending functions, interpolants, or surface constraints. Eiseman presents what he calls "piecewise local control" through the derivation of interpolants and the reader is referred to References 14 and 15 for this approach. Another approach is the construction of control functions which are embedded in the blending functions or surface constraints. Control functions are demonstrated using the two-boundary technique in two dimensions and with cubic blending functions.

The relationship between the computational domain and the physical domain for the two-boundary technique in two dimensions is given by

$$x(\xi,\eta) = x_1(r_1)\mu_1(s) + x_2(r_2)\mu_2(s) + T_1 \frac{dy_1}{dr_1}(r_1)\mu_3(s) + T_2 \frac{dy_2}{dr_2}(r_2)\mu_4(s),$$

$$y(\xi,\eta) = y_1(r_1)\mu_1(s) + y_2(r_2)\mu_2(s) - T_1 \frac{dx}{dr_1}(r_1)\mu_3(s) - T_2 \frac{dx}{dr_2}(r_2)\mu_4(s), \tag{5.1}$$

and

$$\mu_1(s) = 2s^3 - 3s^2 + 1,$$

$$\mu_2(s) = -2s^3 + 3s^2,$$

$$\mu_3(s) = s^3 - 2s^2 + s,$$

$$\mu_4(s) = s^3 - s^2,$$

where

$x_1(r_1),\ y_1(r_1)$ ≡ position on the first boundary as a function of normalized arc length along the boundary

$x_2(r_2),\ y_2(r_2)$ ≡ position on the second boundary as a function of normalized arc length along the boundary

$\frac{dx_1}{dr_1}(r_1),\ \frac{dy_1}{dr_1}(r_1)$ ≡ first derivative along the first boundary with respect to normalized arc length along the boundary

$\frac{dx_2}{dr_1}(r_2),\ \frac{dy_2}{dr_2}(r_2)$ ≡ first derivative along the second boundary with respect to normalized arc length along the boundary

T_1, T_2 ≡ normal derivative magnitudes for the respective boundaries

$r_1 = f_1(\xi)$ ≡ normalized arc length along the first boundary

$r_2 = f_2(\xi)$ ≡ normalized arc length along the second boundary

$s = h[e(\eta)]$ ≡ arc length along the mesh curves connecting the two boundaries

$\left.\begin{array}{l} \varepsilon, \eta \\ 0 \leq \xi \leq 1 \\ 0 \leq \eta \leq 1 \end{array}\right\}$ ≡ coordinates from the computational domain

Uniformly discretizing ξ and η and given the other quantities described above, a corresponding mesh is generated in the physical domain from Eq. (5.1).

For mesh curves connecting the two boundaries (Fig. 8), their relationship and spacing relative to their neighboring mesh curves is based on position, derivatives, and derivative magnitudes at the two boundaries, and the blending functions. Given that the blending functions are the same for all mesh curves, the spacing between the curves is only a function of boundary information. The boundary positions and derivatives are a function of normalized arc lengths which are in turn written as functions of the computational coordinate ξ. It is the functions $f_1(\xi)$ and $f_2(\xi)$ that ultimately control the spacing between mesh curves. When there is relatively low slope in these functions (Fig. 8), there is concentration of

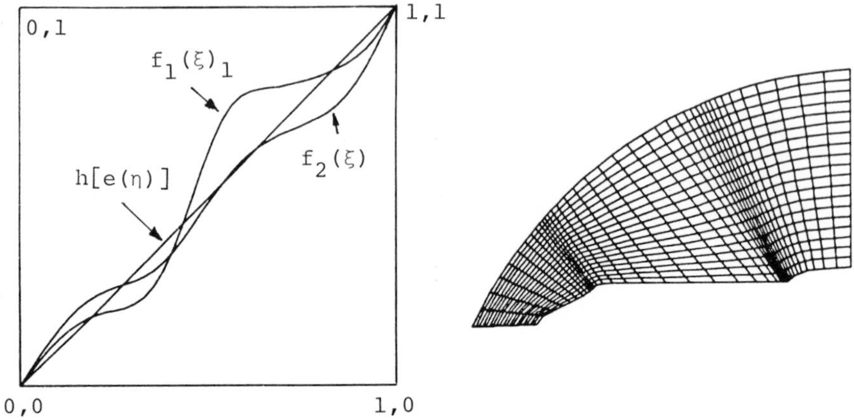

Fig. 8. Effect of control functions.

mesh curves, and when there is relatively high slope the mesh curves are dispersed (Fig. 9). In a similar manner the mesh

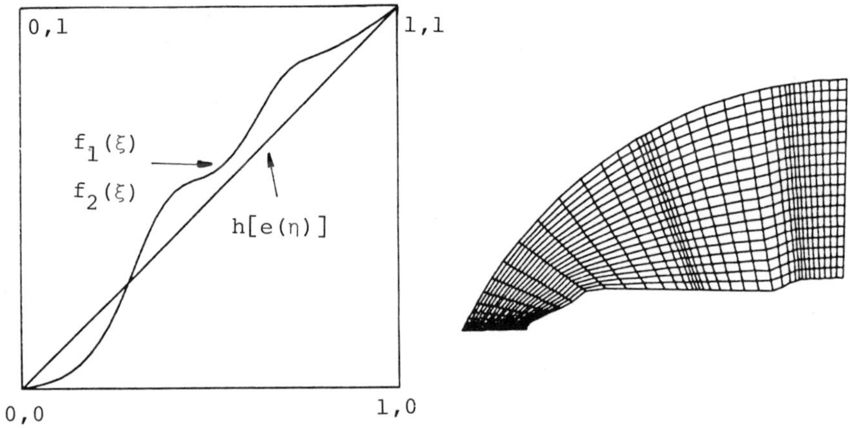

Fig. 9. Spacing between neighboring mesh curves.

points along a mesh curve are distributed by the blending functions. A control function $h[e(\eta)]$ relating η to the normalized arc length determines the final mesh point distribution along the mesh curve (Fig. 10). The functions $f_1(\xi), f_2(\xi)$ and $h[e(\eta)]$ are called control functions. They should be single-valued, smooth, and have smooth derivatives.

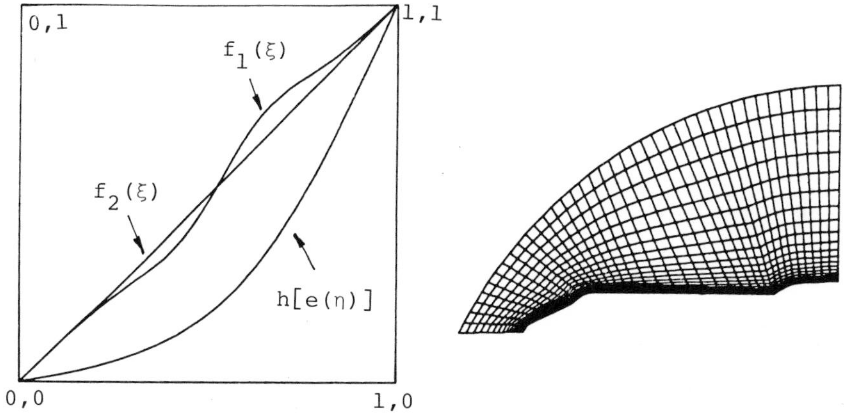

Fig. 10. Control of mesh points along mesh curves.

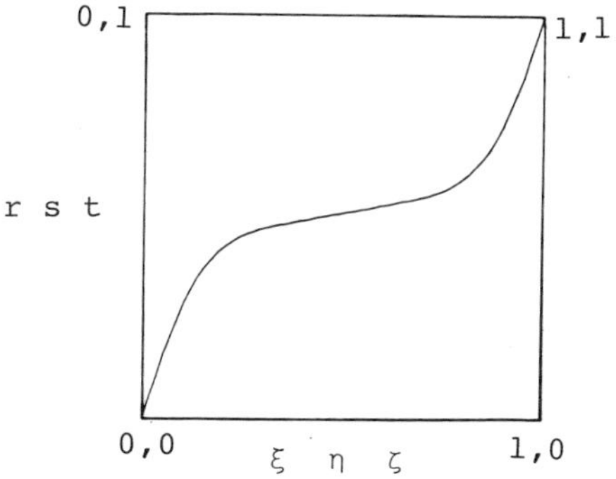

Fig. 11. Domain for the definition of control functions.

Another condition is that the functions are defined on the unit square (Fig. 11). The control functions can be analytic functions such as

$$r_1 = \frac{e^{\hat{K}\xi} - 1}{e^{\hat{K}} - 1}$$

where the parameter $\hat{K}$ controls the concentration of mesh curves near the first boundary. In general, analytic functions are restrictive relative to where control can be applied. Another approach for arbitrary control is the use of smoothing spline functions on the unit square. This approach is described in Reference 17.

6. MESH GENERATION TOPOLOGY.

The algebraic mesh generation techniques that are presented are defined with the assumption that a uniform rectangular computational domain transforms into a physical domain. Also, exterior boundaries of the computational domain transform into boundaries on the physical domain. Consequently the topology of the physical domain strongly influences how a mesh generation technique is applied. It is obvious that a single six-sided box (computational domain) or a square in two dimensions is not going to transform into all physical domains. Further, in certain cases, transformations can only be made by introducing singularities. Problems most often arise when boundaries are closed.

7. COMPUTATIONAL ASPECTS.

The computational requirements for generating a mesh and the Jacobian matrix associated with a mesh using algebraic techniques are relatively minimal. It is necessary, however, to visually assure that desired constraints are satisfied. Ultimately, this phase of problem solving should be in an interactive mode with high band width communications between the computer and a graphics terminal [17]. The Jacobian matrix at every mesh point(transformation data) can be computed once and stored for later use. Alternately, the mesh coordinates can be stored and the transformation data computed when needed by first finite differencing the physical coordinates with respect to the computational coordinates and computing the Jacobian matrix at the mesh points as described in Eq. (2.3). The tradeoff is between additional storage for the first approach or additional computation for the latter.

The computational requirements for large scale viscous-compressible aerodynamic simulation are extreme and tax the capabilities of any presently existing computer [18]. The approach that is described here is to solve the transformed Navier-Stokes equations using a proven explicit numerical technique on a vector computer [19]. The computer is the CDC CYBER 203 where the architecture is based on vector processing with virtual memory storage. In addition to the vector architecture, there are two aspects relative to the computer that are very important for the explicit solution of the Navier-

Stokes equations: (1) the capability of halfword arithmetic and storage; and (2) the effect of data transfer between secondary memory and primary memory. Halfword arithmetic allows for meshes that are twice as large as would be otherwise possible. Frequent transfers of data from primary memory to secondary memory should be minimized or avoided through data management in the solution algorithm or by constraining the mesh size on which a solution is attempted. The transfer of data to and from secondary memory is relatively slow compared to the access of primary storage and consequently can result in a "thrashing" [1] situation.

Another computational aspect is that a "Navier-Stokes solver" can be relatively general. The application of initial and boundary conditions can be performed in separate subroutines from the general solution procedure. Defining a new problem by initial and boundary conditions does not require major program modifications.

In the following section the MacCormack time-split algorithm is briefly described for its application on the CYBER 203 computer. Program organization and how it relates to the virtual memory is presented. Detailed descriptions can be found in References 1 and 2. It should be noted here that the source of the transformation data is independent of the Navier-Stokes solver.

7.1 Computational algorithm.

The computational technique described herein is the MacCormack time-split predictor-corrector algorithm [18-20] which was proposed about 1970 and is a derivative of the MacCormack unsplit predictor-corrector algorithm [21]. Both techniques are explicit which implies that they are time step stability limited and both techniques are second order accurate. An advantage of the MacCormack techniques is that they are relatively easy to apply to the transformed equations of motion (Eq. (3.2)). The split operator algorithm has the added advantage that different time step magnitudes can be used in each operator. A third hybrid scheme [22] can be applied by subdividing the operators into implicit and explicit portions. This approach, however, is more complex and the success of its use is somewhat case dependent. The explicit time-split technique is chosen because of its simplicity and vectorization characteristics for application on the CYBER 203 computer.

The split algorithm consists of a predictor and corrector step for each coordinate direction. Consequently, a predictor and corrector step for a coordinate direction is called an operator for that direction (i.e., $L_{\text{direction}}$ (time step)). A time step is completed in this algorithm with the application of each operator applied symmetrically about the operator for the coordinate direction of primary flow. That is

$$U^{n+1}_{i,j,k} = \left[L_\eta(\Delta t_\eta)\right]\left[L_\zeta(\Delta t_\zeta)\right]\left[L_\xi(\Delta t_\xi)\right]\left[L_\zeta(\Delta t_\zeta)\right]\left[L_\eta(\Delta t_\eta)\right] U^n_{i,j,k}$$

where

$$\Delta t_\eta = \Delta t_\zeta = \frac{1}{2}\,\Delta t_\xi .$$

Each operator is defined by an output state solution U^{out} for a given input state solution U^{in}. For instances

$$L_\xi(\Delta t_\xi) = U^{out}_{i,j,k}$$

where

Predictor step:

$$\bar{U}_{i,j,k} = U^{in}_{i,j,k} - \frac{\Delta t_\xi}{\Delta\xi}\left[(F_i - F_{i-1})\,\frac{\partial\xi}{\partial x}\,i + (G_i - G_{i-1})\,\frac{\partial\xi}{\partial y}\,i + (H_i - H_{i-1})\,\frac{\partial\xi}{\partial z}\,i\right]_{j,k};$$

Corrector step:

$$U^{out}_{i,j,k} = \frac{1}{2}\left(U^{in}_{i,j,k} + \bar{U}_{i,j,k} - \frac{\Delta t_\zeta}{\Delta\zeta}\left[(F_{k+1} - F_k)\,\frac{\partial\zeta}{\partial x}\,k + (G_{k+1} - G_k\,\frac{\partial\zeta}{\partial y}\,k + (H_{k+1} - H_k)\,\frac{\partial\zeta}{\partial z}\,k\right]_{i,j}\right).$$

It has been noted that the MacCormack algorithms are second order accurate. Forward and backward differences are applied such that after the predictor and corrector steps are completed an effective central difference approximation is obtained [1, 2].

The split MacCormack algorithm is time step stability limited, and there is no complete stability analysis to indicate the maximum allowable time step. A conservative time step is

$$\Delta t \leq \min \left[\frac{|u|}{\Delta x} + \frac{|v|}{\Delta y} + \frac{|w|}{\Delta z} + c\sqrt{\frac{1}{\Delta x^2} + \frac{1}{\Delta y^2} + \frac{1}{\Delta z^2}} \right]^{-1}$$

where

$c \equiv$ local speed of sound.

7.2 Application of vector processing to the computational

The MacCormack time-split algorithm has been programmed to run on the CDC STAR-100 and CYBER 203 computers. The program called the Navier-Stokes solver was first written in STAR FORTRAN and is described in Reference 23. The program has since been written in the SL/1 language [24] where 32-bit arithmetic is used to increase the computational speed and incore storage. Thirty-two bit FORTRAN has recently been made available, and it is anticipated that the SL/1 language will be phased out. The split algorithm requires several passes through the data base per time step. When the data base exceeds the primary storage capacity of the CYBER 203 computer a time penalty is imposed when data is called from secondary memory. However, a data management procedure has been implemented to minimize the penalties associated with the use of secondary memory.

The CYBER 203 is a vector processing computer capable of achieving high result rates when a high degree of parallelism is present in the computation. When an identical operation is to be performed on consecutive elements in memory, a vector instruction is issued to perform the operation. Each vector instruction involves a time penalty, called vector startup, regardless of the length of the vector. As the length of the vector increases, the operation becomes more efficient since the penalty becomes relatively less important.

The CYBER 203 has about one million words of primary memory with virtual memory architecture. Memory is referred to as pages. The two page sizes on the CYBER 203 are "small" pages which are 512 64-bit words and "large" pages which are

65536 words or 128 small pages. A user can have access to about 15 large pages in primary memory at any one time. The movement of data from secondary memory into primary memory involves moving pages of data in and out of primary memory. This is called a "page fault" and involves a startup time and transmission time similar to vector operations. It then becomes important to make the most efficient use of the data when it is in primary memory in order to avoid a situation where the machine is spending more time moving pages of data in and out of primary memory than it is spending on actual computations. This is often referred to as "thrashing." Storing the data for a large data base program, such as a three-dimensional Navier-Stokes solver, in a conventional manner could very possibly lead to this situation. If, however, you design an interleaved data base [1] where the variables that are currently being used are stored together then it could result in less movement of "pages" of data.

Normally it may be thought that vector lengths should be equal to the total number of mesh points and vector operations sweep through the entire mesh with each variable dimensioned to the number of mesh points. However, for the number of variables involved and the large number of mesh points this leads very quickly to the "thrashing" situation described earlier. Instead, vectors are computed in planes in the ξ directions (Fig. 12) with vector lengths approximately equal to the number of mesh points in a plane. Temporary reusable vectors are maintained for three local planes and a four-dimensional array S(I,L,J,K) contains the five state variables and nine elements of the Jacobian matrix for each mesh point.

In order to minimize the sweeps through the data base S(I,L,J,K), the corrector step for a plane is performed as soon as enough planes of the predictor step are available (Fig. 13). Consequently for the application of each operator, there is one sweep through the data base and five sweeps for a time step.

Within a ξ plane, a vector sweep is from the lower left hand corner to the upper right hand corner. The exact length, starting point, and end point of the vector is dependent on

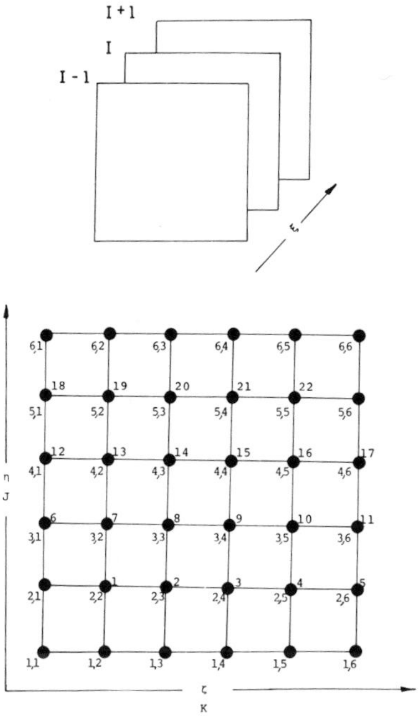

Fig. 12. Vector arrangement in planes.

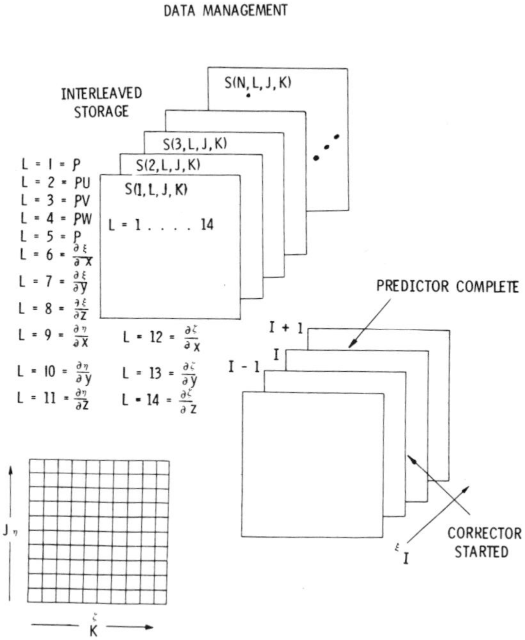

Fig. 13. Data management for Navier-Stokes solver.

the operator and the direction of the finite differencing. Vector operations include boundary points where erroneous values are computed during a vector computation. The boundary condition subroutine is called to compute the boundary conditions and overwrite the erroneous values.

The transformation data which consist of the nine derivatives of the computational coordinates with respect to the physical coordinates at each mesh point are computed in a separate program and stored on a disk file. Once a geometry is established, the transformation data remains constant.

8. APPLICATIONS.

The algebraic mesh generation procedures and the Navier-Stokes solver have been used to solve several large-scale viscous-compressible flow problems. Three examples are found in References 25, 26, and 27. The Mach 7 flow of air over a spherical protuberance on a flat plate [27] is briefly described. In this case only five of the nine elements of the Jacobian matrix are nonzero, and this allows the number of mesh points in the Navier-Stokes solver to be increased to 78208 points. The plate surface mesh, the centerline windward surface mesh, and a centerline crossflow surface mesh are shown in an exploded view in Figure 14. The mesh spacing in

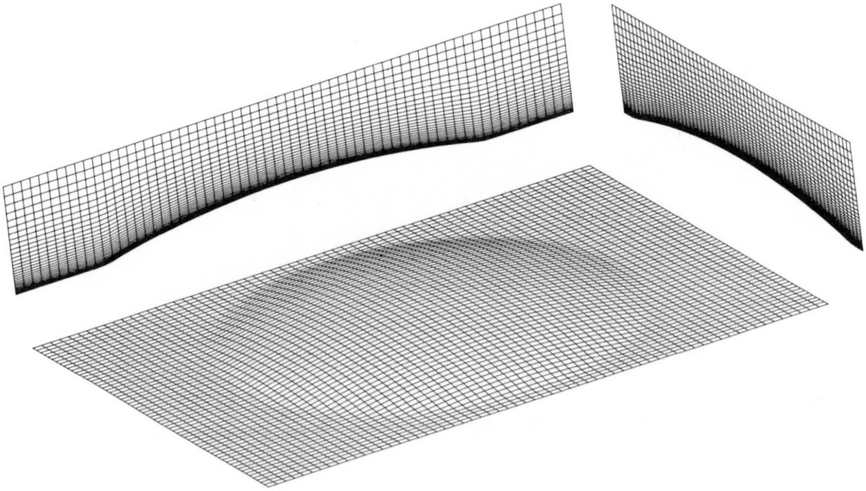

Fig. 14. Exploded view of protuberance mesh.

the y coordinate direction is exponentially increased away from the solid surface through a simple application of the two-boundary technique. The initial inflow is obtained from a boundary-layer solution and the Navier-Stokes solver produces the downstream solution. Figures 15 and 16 show the pressure distribution and temperature gradient distribution. For further details the reader is referred to Reference 27.

Fig. 15. Pressure distribution on protuberance mesh.

Fig. 16. Temperature gradient on protuberance surface.

9. CONCLUSIONS.

Large-scale aerodynamic simulation using finite difference techniques implies that a large number of mesh points are strategically and orderly placed in the domain of the flow field. The algebraic procedures for mesh generation described herein have proven to be viable for two-dimensional meshes and three-dimensional meshes where the meshes are defined in a plane. Also, the algebraic techniques are potentially the most satisfactory approach to arbitrary three-dimensional meshes where mesh points are not confined to planes.

Viscous-compressible aerodynamic simulation is the solution of the Navier-Stokes equations. For the application of mesh generation which accounts for boundary fitting and capture of high gradient phenomena with a limited number of mesh points, the equations are transformed to an idealized computational coordinate system. A description of the solution of the transformed equations using a time-split MacCormack technique on a vector computer has been presented. A relationship between mesh generation, simulation procedure, and advanced computational equipment has been shown.

REFERENCES

1. Smith, R. E., Two-Boundary Grid Generation for the Solution of the Three-Dimensional Navier-Stokes Equations, Ph. D. Dissertation, Old Dominion University, 1981.
2. Smith, R. E., Two-Boundary Grid Generation for the Solution of the Three-Dimensional Navier-Stokes Equations, NASA TM-83123, May, 1981.
3. Thompson, J. F., F. C. Thames, and C. W. Mastin, Automatic Numerical Generation of Body-Fitted Curvilinear Coordinate Systems for Fields Containing Any Number of Arbitrary Two-Dimensional Bodies, Journal of Computational Physics, Vol. 15, July, 1974, pp. 299-319.
4. Thames, F. C., J. F. Thompson, C. W. Mastin, and R. L. Walker, Numerical Solution for Viscous and Potential Flow About Arbitrary Two-Dimensional Bodies Using Body-Fitted Coordinate Systems, Journal of Computational Physics, Vol. 24, July, 1977, pp. 245-273.

5. Thompson, J. F., F. C. Thames, and S. P. Shanks, Use of Numerically Generated Body-Fitted Coordinate Systems for Solutions of the Navier-Stokes Equations, Proceedings of the AIAA 2nd Computational Fluid Dynamics Conference, Hartford, CT, July, 1975.
6. Smith, R. E., Algebraic Grid Generation, Numerical Grid Generation, J. F. Thompson, ed., Elsevier Sciences Publishing Co., 1982.
7. Steger, J. L., Implicit Finite Difference Simulation of Flow About Arbitrary Geometries with Applications to Airfoils, AIAA paper 80-0192, AIAA 18th Aerospace Sciences Meeting, Jan., 1980, Pasadena, California.
8. Hindman, R. G., Geometrically Induced Errors and Their Relationship to the Form of the Governing Equations and the Treatment of Generalized Mappings, AIAA paper 81-1008, AIAA 5th Computational Fluid Dynamics Conference, Palo Alto, California, June, 1981.
9. Gordon, W. J., and C. A. Hall, Construction of Curvilinear Coordinate Systems and Application to Mesh Generation, International Journal for Numerical Methods in Engineering, Vol. 7, 1973, pp. 461-477.
10. Eriksson, Lars-Erik, Three-Dimensional Spline-Generated Coordinate Transformations for Grids Around Wing-Body Configurations, Numerical Grid Generation Techniques, NASA CP 2166, 1980.
11. Rizzi, A., and L. E. Eriksson, Transfinite Mesh Generation and Damped Euler Equation Algorithm for Transonic Flow Around Wing-Body Configurations, AIAA 5th Computational Fluid Dynamics Conference, Palo Alto, California, June, 1981.
12. Eiseman, P. R., A Multi-Surface Method of Coordinate Generation, Journal of Computational Physics, Vol. 33, No. 1, Oct., 1979.
13. Eiseman, P. R., Geometric Methods in Computational Fluid Dynamics, ICASE Report No. 81-11, April, 1980.
14. Eiseman, P. R., A Multi-Surface Method for Coordinate Generation, Journal of Computational Physics, Vol. 29, 1978.

15. Eiseman, P. R., Three-Dimensional Coordinates About Wings, 4th AIAA Computer Fluid Dynamics Conference, Williamsburg, Virginia, July, 1979.
16. Smith, R. E., and B. L. Weigel, Analytic and Approximate Boundary-Fitted Coordinate Systems for Fluid Flow Simulation, AIAA paper 80-0192, AIAA 18th Aerospace Sciences Meeting, Pasadena, California, January, 1980.
17. Smith, R. E., R. A. Kudlinski, and E. L. Everton, A Grid Spacing Control Technique for Algebraic Grid Generation Methods, AIAA paper 82-0226, AIAA 20th Aerospace Sciences Meeting, Orlando, Florida, January, 1982.
18. MacCormack, R. W., and A. J. Paullay, Computational Efficiency Achieved by Time Splitting of Finite Difference Operators, AIAA paper 72-154, January, 1972.
19. Shang, J. S., and W. L. Hankey, Numerical Solution of the Navier-Stokes Equations for a Three-Dimensional Corner, AIAA paper 77-169, Los Angeles, CA, also AIAA Journal, Vol. 15, Nov., 1977, pp. 1575-82.
20. Shang, J. S., W. L. Hankey, and J. S. Petty, Three-Dimensional Supersonic Interacting Turbulent Flow Along a Corner, AIAA paper 78-1210, Seattle WA, July, 1978.
21. MacCormack, R. W., The Effect of Viscosity in Hypervelocity Impact Cratering, AIAA paper 69-354, May, 1969.
22. MacCormack, R. W., An Efficient Numerical Method for Solving the Time Dependent Compressible Navier-Stokes Equations at High Reynolds Number, Computing in Applied Mechanics, ADM, Vol. 18, New York Society of Mechanical Engineering.
23. Smith, R. E., and J. I. Pitts, The Solution of the Three-Dimensional Compressible Navier-Stokes Equations on a Vector Computer, Third IMAC International Symposium on Computer Methods for Partial Differential Equations, Lehigh University, PA, June, 1979.
24. SL/1 Reference Manual, Analysis and Computation Division, NASA Langley Research Center, Hampton, VA.

25. Smith, R. E., Numerical Solution of the Navier-Stokes Equations for a Family of Three-Dimensional Corner Geometries, AIAA paper 80-1349, AIAA 13th Fluid and Plasma Dynamics Conference, Snowmass, Colorado, July, 1980.

26. Shang, J. S., W. L. Hankey, and R. E. Smith, Flow Oscillations of Spike-Tipped Bodies, AIAA paper 80-0062, also AIAA Journal, Vol. 20, No. 1, January, 1982.

27. Olsen, G. C., and R. E. Smith, Analysis of Aerothermal Loads on Spherical Dome Protuberances, AIAA paper 83-1557, AIAA 18th Thermophysics Conference, Montreal, Canada, June, 1983.

National Aeronautics & Space Admin
Langley Research Center
Hampton, VA 23665

SOFTWARE FOR THE SPECTRAL ANALYSIS OF SCALAR AND VECTOR FUNCTIONS ON THE SPHERE

Paul N. Swarztrauber

1. INTRODUCTION

In this paper we describe several software packages that can be used to analyze both scalar and vector functions on the sphere. The methods that have been implemented in the software are also discussed and indeed they occupy the major part of the paper. The methods are discussed at the beginning of each section and the software is presented at the end.

The analysis of scalar functions on the sphere is given in the next section. This analysis is significantly different from the analysis on the plane and in Cartesian coordinates. For a doubly periodic tabulation $f_{i,j}$ defined on the plane, the spectral analysis consists of determining coefficients $a_{m,n}$ such that a linear combination of complex exponentials interpolates the data. That is, we wish to find $a_{m,n}$ such that

$$f_{i,j} = \sum_n \sum_m a_{m,n} e^{i(mx_i+ny_j)} \tag{1.1}$$

for all i and j. It is known that this interpolation problem has a solution, or equivalently that the coefficients $a_{m,n}$ can be found such that the series in (1.1) passes through each of the points $f_{i,j}$. The coefficients $a_{m,n}$ constitute the discrete spectral analysis of $f_{i,j}$.

ISBN 0-12-546080-5

The main observation in the next section is that the interpolation problem on the sphere does not have a solution. Given a tabulation $f_{i,j}$ on a uniform latitude-longitude grid, it is not always possible to find a linear combination of the spherical harmonics that passes through each $f_{i,j}$. Nevertheless, analyses can be performed on the sphere and the coefficients are determined as the solution to a weighted least-squares problem.

A discrete set of basis functions is selected as a finite number of spherical harmonics that do not alias on the grid. The analysis consists of determining a linear combination of the discrete basis functions that approximates the $f_{i,j}$ in a weighted least-squares sense. Although there are a number of ways [8] in which to determine these coefficients, we use the method given by Machenhauer and Daley [5]. Their analysis is unique in the sense that it is exact for any tabulation $f_{i,j}$ of a function that can be expressed in terms of the discrete basis. The key to this analysis is certain functions $Z_n^m(\theta)$ that play a role in the analysis similar to that played by the associated Legendre functions $P_n^m(\theta)$ in the synthesis. The $Z_n^m(\theta)$ functions also determine the weights and the norm associated with the weighted least-squares problem.

At the end of Section 2 there is a brief discussion of subroutines SHA and SHS for the analysis and synthesis of discrete scalar functions on the sphere. The analysis of vector functions in terms of the vector spherical harmonics is given in Section 3. Unlike analysis on the plane in Cartesian coordinates, analysis of a vector function on the sphere cannot be performed in terms of components using the spherical harmonics because the components of a vector function are likely to be discontinuous or at least to have discontinuous derivatives at the poles. These discontinuities are induced by the spherical coordinate system since the same vector function can have smooth components in Cartesian coordinates. Since the spherical harmonics are smooth at the poles they do not provide a suitable set of basis functions for the components of the vector function.

The vector spherical harmonics [6] provide a suitable set of vector basis functions. Like vector functions in general, the vector spherical harmonics include functions that are discontinuous at the poles. Any vector function, even with discontinuous components, can be uniformly approximated by a linear combination of spherical vector harmonics if its components are smooth in Cartesian coordinates. Subroutines VSHA and VSHS for the vector spherical harmonic analysis and synthesis are described at the end of the third section.

The use of spectral methods for solving partial differential equations (PDEs) on the sphere is particularly attractive since many of the terms in the PDEs are unbounded in the neighborhood of the poles. Each of these unbounded terms can be included with certain other unbounded terms which cancel in such a way as to produce a bounded expression. These bounded expressions together with their unbounded terms are given in [9] for all first- and second-order differential expressions. One can avoid evaluating the unbounded terms by using the spectral method in which only the bounded expressions are applied to the spectral representation of the vector function. For this reason, the spectral method eliminates the "polar problem" that is associated with finite difference models of geophysical processes on the sphere.

In Section 4 we describe the spectral method for computing the modes of a specific vector differential equation. In particular, we outline the analysis of the linearized shallow-water equations or Laplace tidal equations. The details of this analysis are given in [10]. The modes or Hough functions are expressed as a series in the vector spherical harmonics. The frequencies and coefficients in the series are computed as the eigenvalues and eigenvectors of an infinite symmetric pentadiagonal matrix. A software package for computing the Hough functions, called HOUGHPACK, is described at the end of Section 4.

In Section 5 we describe two utility packages. The first, called ALFPACK, contains 16 programs for computing the associated Legendre functions $P_n^m(\theta)$ and approximately 3,000 lines of code. The second package, called FFTPACK, contains 19 programs for computing various fast Fourier transforms and about 3,000 lines of code.

2. THE SPECTRAL ANALYSIS OF SCALAR FUNCTIONS

Spectral analysis on the rectangle and in Cartesian coordinates is very straightforward. The analysis is obtained by solving the interpolation problem. Given the tabulation $f_{i,j}$ at grid points (x_i, y_j) where $x_i = i\frac{2\pi}{M}$ and $y_j = j\frac{2\pi}{N}$ then there exist coefficient $a_{m,n}$ such that

$$f(x,y) = \sum_n \sum_m a_{m,n} e^{i(mx+ny)} \tag{2.1}$$

interpolates the tabulation, i.e., such that

$$f(x_i,y_i) = f_{i,j}. \tag{2.2}$$

The coefficients $a_{m,n}$ provide the spectral analysis of the tabulation $f_{i,j}$.

Spectral analysis on the surface of the sphere is quite different from analysis on the rectangle because the interpolation problem on the sphere does <u>not</u> have a solution. Given a tabulation $f_{i,j}$, at grid points (θ_i,ϕ_j) in spherical coordinates where $\theta_i=i\frac{\pi}{M}$ and $\phi_j=\frac{2\pi}{N}$, then coefficients $a_{m,n}$ <u>cannot</u> necessarily be found such that the function

$$f(\theta,\phi)=\sum_n \sum_m a_{m,n} Y_n^m(\theta,\phi) \tag{2.3}$$

interpolates $f_{i,j}$, i.e., such that $f(\theta_i,\phi_j)=f_{i,j}$ for all i and j. The $Y_n^m(\theta,\phi)$ are the spherical harmonics given by

$$Y_n^m(\theta,\phi) = P_n^m(\theta)\, e^{im\phi} \tag{2.4}$$

where the $P_n^m(\theta)$ are the associated Legendre functions,

$$P_n^m(\theta) = \frac{1}{2^n n!} (\sin\theta)^m \frac{d^{m+n}}{dx^{m+n}} (x^2-1)^n \tag{2.5}$$

and $x = \cos\theta$.

Any trivariate polynomial $p(x,y,z)$ defined on the surface of the sphere can be expressed as a finite linear combination of spherical harmonics. Therefore the interpolation problem with polynomial basis functions does not have a solution as well. These results are applicable to problems posed on the disk in polar coordinates or more generally to any coordinate system in which the grid points cluster. Spectral analysis on the sphere is obtained as the solution of a weighted least-squares problem rather than the solution of the interpolation problem. There is an attractive aspect to this, namely, that all waves are uniformly resolved on the sphere. Since the grid points cluster near the poles, any computational or observational errors will result in artificially induced high frequencies. These frequencies cannot be resolved by the spherical harmonics analysis and are therefore removed in a uniform way by performing an analysis followed by a synthesis.

In order to motivate the least-squares approach we will first prove that the interpolation problem does not have a solution. The statement of the interpolation problem must include the definition of a finite set of basis functions called the discrete basis. On the rectangle the discrete basis consists of the trigonometric polynomials, of lowest order, that do not alias. By this we mean that any two functions spanned by the set cannot have the same value at all grid points. Since

$$e^{i(n+N/2)y_j} = e^{i(n-N/2+N)j\frac{2\pi}{N}} = e^{i(n-N/2)y_j}$$

the discrete basis includes only those complex exponentials with $|n| < N/2$ and $|m| < M/2$.

The discrete basis for the analysis on the sphere is selected as a finite subset of the spherical harmonics. Since the complex exponential $e^{im\phi}$ occurs in the definition of $Y_n^m(\theta,\phi)$ the discrete basis can include only those spherical harmonics that satisfy $|m| < M/2$.

It is known [1] that the associated Legendre functions have the following form

$$P_n^m(\theta) = \begin{cases} \sum_{k=0}^{n} a_{n,m}(k)\cos k\theta & m \text{ even} \\ \sum_{k=1}^{n} a_{n,m}(k)\sin k\theta & m \text{ odd} \end{cases} \tag{2.6}$$

where $a_{m,n}(k) = 0$ if n-k is odd. For $n > N$ some of the terms in (2.6) will alias. For example,

$$\cos(n+N)\theta_i = \cos(n-N+2N)i\frac{\pi}{N} = \cos(n-N)\theta_i.$$

Thus $Y_n^m(\theta,\phi)$ is included in the discrete basis only if $n \leq N$. The actual definition of the discrete basis as given in [8] is somewhat more detailed and depends on whether m is an even or odd integer.

To show that the interpolation problem does not have a solution it is sufficient to show that the number of discrete basis functions is less than the number of tabulation points. In fact the number of discrete basis functions is about half the number of grid points on the sphere. To simplify the exposition, we will assume that $M = 2N$ and the tabulation $f_{i,j}$ is real. For $m < M/2 = N$ and $n < N$, the discrete basis includes the functions $P_n^m(\theta)\cos m\phi$ and $P_n^m(\theta)\sin m\phi$. The total number of basis functions is about N^2 since the $Y_n^m(\theta,\phi)$ are defined only for $m \leq n$. On the other hand, the number of grid points or tabulated values $f_{i,j}$ is about $MN = 2N^2$. Since there are about twice as many grid points as there are basis functions, we cannot expect to find a linear combination of the functions that will pass through all the points $f_{i,j}$. Thus we conclude that the interpolation problem does not have a solution.

This would seem to contradict the following observation. If we select the complex exponentials as a basis function, then we can find coefficients $a_{m,n}$ such that the function

$$f(\theta,\phi) = \Sigma\,\Sigma\, a_{m,n}e^{i(m\phi+n\theta)} \tag{2.7}$$

does interpolate the tabulation $f_{i,j}$; i.e., $f(\theta_i,\phi_j) = f_{i,j}$ at all grid points. The problem with this approach is that the discrete basis includes functions that are not smooth at the poles. Many of the complex exponentials are not continuous at the poles or have discontinuous derivatives. For example, $e^{i\phi}$ is multivalued at the poles.

Spectral analysis on the sphere must be determined as the solution of a least-squares problem rather than the solution of an interpolation problem. This seems to be an attractive alternative since the linear least-squares problem is well understood. However, if we use the ℓ_2 norm, then the resulting analysis would not share an important property with the analysis on the rectangle, namely that the discrete analysis be exact for any function $f(\theta,\phi)$ that can be expressed as a linear combination of the discrete basis functions. We will require the discrete analysis to be the same as the continuous or exact analysis for any function spanned by the discrete basis. Once a discrete analysis is determined with this property then the correct norm can be determined.

Such an analysis is given by Machenhauer and Daley [5]. For m even define

$$Z_n^m(\theta) = \frac{2}{N} \sum_{k=0}^{N} \cos k\theta \int_0^{\pi} \cos(k\tau) P_n^m(\tau) \sin\tau d\tau \tag{2.8}$$

and for m odd define

$$Z_n^m(\theta) = \frac{2}{N} \sum_{k=0}^{N} \sin k\theta \int_0^{\pi} \sin(k\tau) P_n^m(\tau) \sin\tau d\tau. \tag{2.9}$$

Then the analysis is given by

$$a_{m,n} = \sum_{i=1}^{N} \sum_{j=1}^{M} f_{i,j} \; Z_n^m(\theta_i) e^{im\phi_j} \tag{2.10}$$

and the synthesis is given by

$$\tilde{f}_{i,j} = \sum_{n=0}^{N} \sum_{m=0}^{n} a_{m,n} \; P_n^m(\theta_i) e^{im\phi_j}. \tag{2.11}$$

The tabulations $\tilde{f}_{i,j}$ and $f_{i,j}$ are not necessarily the same since the analysis results from the solution of a least-squares problem. However, the analysis is exact and $\tilde{f}_{i,j} = f_{i,j}$ if the tabulation $f_{i,j}$ can be expressed in terms of the discrete basis.

Let μ be the number of grid points and ν be the number of discrete basis functions. Then the analysis can be written in the matrix form

$$\mathbf{a} = \mathbf{ZQf}. \tag{2.12}$$

The ν vector $\mathbf{a}$ contains the coefficients $a_{m,n}$, and the $\nu \times \mu$ matrix $\mathbf{Z}$ contains the elements $Z_n^m(\theta_i)$. $\mathbf{Q}$ is the $\mu \times \mu$ matrix that corresponds to the Fourier transform in the longitude direction ϕ. The μ vector $\mathbf{f}$ contains the tabulation $f_{i,j}$. The synthesis has the matrix form

$$\mathbf{f} = \mathbf{Q}^T\ \mathbf{Pa} \tag{2.13}$$

where $\mathbf{P}$ is the $\mu \times \nu$ matrix with elements $P_n^m(\theta_i)$. In [8] it is shown that the coefficient vector $\mathbf{a}$ given by (2.12) is the least-squares solution of (2.13) under the norm

$$\|\mathbf{f}\|_s^2 = \|\mathbf{ZQf}\|_2^2 + \|(\mathbf{I} - \mathbf{Q}^T\mathbf{PZQ})\|_2^2. \tag{2.14}$$

It is also shown that $\|\mathbf{f}\|_s$ is consistent with the continuous norm

$$\|f(\theta,\phi)\|_c^2 = \frac{1}{2\pi}\int_0^{2\pi}\int_0^{\pi} f(\theta,\phi)^2 \sin\theta\, d\theta\, d\phi \tag{2.15}$$

in the sense that $\|f(\theta,\phi)\|_c = \|\mathbf{f}\|_s$ for any $f(\theta,\phi)$ that is a linear combination of the discrete basis functions.

The software for the analysis of scalar functions includes two main subroutines. The first is subroutine SHA for the spherical harmonic analysis and the second is subroutine SHS for the spherical harmonic synthesis.

a) SUBROUTINE SHA (L,G,IDIML,A,B,MDIMAB,W)

This subroutine computes the coefficients $a_{m,n}$ and $b_{m,n}$ such that the input array $g_{i,j}$ has the approximate expansion

$$g_{i,j} \cong \sum_{n=0}^{L-1} \sum_{m=0}^{n} P_n^m(\theta_i)(a_{m,n} \cos m \frac{\pi}{\ell-1} + b_{m,n} \sin m \frac{\pi}{\ell-1}). \tag{2.16}$$

The grid is defined by $\theta_i = (i-1) \frac{\pi}{(L-1)}$ for $i = 1, \ldots, L$ and $\phi_i = (j - 1) \frac{\pi}{L-1}$ for $j = 1, \ldots, 2(L-1)$. Arrays $a_{m,n}$, $b_{m,n}$ and $g_{i,j}$ are stored in A(M,N), B(M,N) and G(I,J), respectively. IDIMG is the first dimension of array G and MDIMAB is the first dimension of arrays A and B. W is a work array.

b) SUBROUTINE SHS (L,G,IDIMG,A,B,MDIMAB,W)

This subroutine computes the spherical harmonic synthesis. Given the coefficients $a_{m,n}$ and $b_{m,n}$, then SHS computes $g_{i,j}$ from (2.16).

3. THE SPECTRAL ANALYSIS OF VECTOR FUNCTIONS

Unlike the analysis on a rectangle, the scalar analysis on the sphere cannot be applied to the components of a vector function. This is due to the fact that the components of a vector function on the sphere are likely to be discontinuous or to have discontinuous derivatives at the poles. For example, consider the vector function defined by the velocity components of a unit sphere in ridged rotation about the equator. That is, the axis of rotation is perpendicular to the coordinate axis z and coincident with the y axis.

In Cartesian coordinates the velocity components are $v_x = z$, $v_y = 0$, and $v_z = -x$, which are continuously differentiable everywhere as functions of x, y, and z. However, the same vector function has components $v_r = 0$, $v_\theta = \cos\phi$, and $v_\phi = \cos\theta\sin\phi$ in spherical coordinates. Both v_θ and v_ϕ are discontinuous (multivalued) at the poles $\theta = 0, \pi$. The value of v_θ at $\theta = 0$ depends on the value of ϕ along which the pole is approached. These discontinuities in the velocity components are, of course, not physical but rather are induced by the discontinuities in the spherical coordinate system.

The multivalued character of vector functions at the pole creates a number of computational problems for models of geophysical processes on the sphere. While the vector function is itself multivalued, many of its derivatives are unbounded at the poles. As a result, many of the terms in a partial differential equation in spherical coordinates will be unbounded in the neighborhood of the poles. For example, the viscous terms in the Navier-Stokes equations contain the terms

$$E\mathbf{v} = \frac{1}{\sin^2\theta}\frac{\partial^2 v_\theta}{\partial\phi^2} - \frac{2\cos\phi}{\sin^2\theta}\frac{\partial v_\phi}{\partial\phi} - \frac{v_\theta}{\sin^2\theta} . \tag{3.1}$$

Now let $v_\theta = \cos\phi$ and $v_\phi = -\cos\theta\sin\phi$ be the components given above that correspond to rigid rotation, then

$$E\mathbf{v} = -\frac{\cos\phi}{\sin^2\theta} + 2\frac{\cos^2\theta\cos\phi}{\sin^2\theta} - \frac{\cos\phi}{\sin^2\theta} = -2\cos\phi . \tag{3.2}$$

It is interesting to note that although the expression is bounded, its individual terms are unbounded at the poles. All of the unbounded terms in a partial differential equation can be included in a bounded expression. A complete list of bounded expressions with unbounded terms is given in [9]. As a result of the foregoing considerations it becomes evident that any set of basis functions for vector functions on the sphere must include multivalued functions with unbounded derivatives. The spherical harmonics are not a suitable basis since they are smooth at the poles.

In order to find a suitable basis, we begin with the Helmholtz relations. For any vector function $\mathbf{v} = (0, v_\theta, v_\phi)$ with zero radial component, there exist scalar functions Φ and Ψ such that

$$v_\theta = \frac{\partial\Phi}{\partial\theta} + \frac{1}{\sin\theta}\frac{\partial\Psi}{\partial\phi} \tag{3.3}$$

$$v_\phi = \frac{1}{\sin\theta}\frac{\partial\Phi}{\partial\phi} - \frac{\partial\Psi}{\partial\theta} \tag{3.4}$$

where Φ and Ψ are scalar functions that can be represented in terms of spherical harmonics.

A sequence of vector functions can be generated by substituting the spherical harmonics into the right side of the Helmholtz relations. It seems reasonable to consider the set of vector functions that is obtained by first substituting $\Phi = Y_n^m$ and $\Psi = 0$ and then $\Phi = 0$ and $\Psi = Y_n^m$ as candidates for a basis for vector function. In doing so we in fact obtain the horizontal components of the spherical vector harmonics $\mathbf{B}_{m,n}$ and $\mathbf{C}_{m,n}$ as given in [6]. It can be shown that the radial component of a smooth vector function is also smooth and thus representable in terms of the spherical harmonics. For $n = 1,2,...$ and $m = -n,...,$ n the vector spherical harmonics are given by

$$\mathbf{P}_{m,n} = \begin{bmatrix} P^{n}_{m} \\ 0 \\ 0 \end{bmatrix} e^{im\phi}, \quad \mathbf{B}_{m,n} = \begin{bmatrix} 0 \\ A_n^m \\ iB_n^m \end{bmatrix} \frac{e^{im\phi}}{\sqrt{n(n+1)}},$$

$$\mathbf{C}_{m,n} = \begin{bmatrix} 0 \\ iB_n^m \\ -A_n^m \end{bmatrix} \frac{e^{im\phi}}{\sqrt{n(n+1)}} \tag{3.5}$$

where A_n^m and B_n^m are functions of θ only

$$A_n^m = \frac{d}{d\theta} P_n^m = \frac{1}{2}[(n+m)(n-m+1)P_n^{m-1} - P_n^{m+1}] \tag{3.6}$$

and

$$B_n^m = \frac{m}{\sin\theta} P_n^m = \frac{1}{2}[(n+m)(n+m-1)P_{n-1}^{m-1} + P_{n-1}^{m+1}] \tag{3.7}$$

The vectors $\mathbf{B}_{m,n}$ and $\mathbf{C}_{m,n}$ have discontinuities at the poles that "match" those of the vector functions that we wish to represent. The spherical vector harmonics are orthogonal under the inner product

$$(\mathbf{u},\mathbf{v}) = \int_0^{2\pi} \int_0^{\pi} \mathbf{u}^*\mathbf{v} \sin\theta d\theta d\phi$$

where $\mathbf{u}$ and $\mathbf{v}$ are vector functions on the sphere and $\mathbf{u}^*$ is the conjugate transpose of $\mathbf{u}$. If $j \neq m$ then

$$(\mathbf{B}_{m,n}, \mathbf{C}_{j,k}) = 0$$

since the complex exponentials $e^{im\phi}$ and $e^{ij\phi}$ are orthogonal. It is also known that

$$(\mathbf{P}_{m,n}, \mathbf{P}_{m,k}) = (\mathbf{B}_{m,n}, \mathbf{B}_{m,k}) = (\mathbf{C}_{m,n}, \mathbf{C}_{m,k})$$

$$= \begin{cases} \dfrac{4\pi}{2n+1} \dfrac{(n+m)!}{(n-m)!} & \text{if } n = k \\ 0 & \text{if } n \neq k \end{cases} \tag{3.8}$$

The spectral analysis of a vector function $\mathbf{f}(\theta,\phi)$ consists of determining coefficients $p_{m,n}$, $b_{m,n}$, and $c_{m,n}$ such that:

$$\mathbf{f}(\theta,\phi) = \sum_n \sum_m (p_{m,n} \mathbf{P}_{m,n} + b_{m,n} \mathbf{B}_{m,n} + c_{m,n} \mathbf{C}_{m,n}). \tag{3.9}$$

Since the vector spherical harmonics are orthogonal, the coefficients are given formally as

$$p_{m,n} = \frac{2n+1}{4\pi} \frac{(n-m)!}{(n+m)!} \int_0^{2\pi} \int_0^{\pi} \mathbf{P}^*_{m,n} \mathbf{f}(\theta,\phi) \sin\theta d\theta d\phi \tag{3.10}$$

$$b_{m,n} = \frac{2n+1}{4\pi} \frac{(n-m)!}{(n+m)!} \int_0^{2\pi} \int_0^{\pi} \mathbf{B}^*_{m,n} \mathbf{f}(\theta,\phi) \sin\theta d\theta d\phi \tag{3.11}$$

$$c_{m,n} = \frac{2n+1}{4\pi} \frac{(n-m)!}{(n+m)!} \int_0^{2\pi} \int_0^{\pi} \mathbf{C}^*_{m,n} \mathbf{f}(\theta,\phi) \sin\theta d\theta d\phi \tag{3.12}$$

Even though $\mathbf{f}(\theta,\phi)$, $\mathbf{B}_{m,n}$, and $\mathbf{C}_{m,n}$ are multivalued at the poles, with unbounded derivatives, it is known that the series converges uniformly for any vector function $\mathbf{f}(\theta,\phi)$ with components that are smooth in Cartesian coordinates.

The discrete analysis of vector function parallels that of the discrete analysis of scalar functions.

a) The discrete vector basis is selected as the finite subset of $\mathbf{P}_{m,n}$, $\mathbf{B}_{m,n}$ and $\mathbf{C}_{m,n}$ whose components do not alias on the grid in spherical coordinates.

b) The discrete vector analysis is constructed in such a way as to be exact for any vector function that is a linear combination of the finite discrete vector basis.

c) Like the scalar analysis, the vector analysis is obtained as the solution of a least-squares problem.

The software for the analysis of vector functions includes two main subroutines. The first is subroutine VSHA for the vector spherical harmonic analysis of the horizontal components of the vector function. The second is subroutine VSHS which computes the vector spherical harmonic synthesis.

a) SUBROUTINE VSHA (L,IDWV,V,W,MDAB,A1,B1,A2,B2,WORK)

Given the colatitudinal and longitudinal components $v_{i,j}$ and $w_{i,j}$, respectively, VSHA computes coefficients $a1_{m,n}$, $b1_{m,n}$, $a2_{m,n}$, and $b2_{m,n}$ such that

$$\begin{bmatrix} v_{ij} \\ w_{ij} \end{bmatrix} \cong \sum_{n=0}^{L-1} \sum_{m=0}^{n} (a1_{m,n} + ib1_{m,n}) \mathbf{B}_{m,n} + (a2_{m,n} + ib2_{m,n}) \mathbf{C}_{m,n}. \tag{3.13}$$

The grid is defined by $\theta_i = (i-1) \frac{\pi}{(L-1)}$ for i=1,...,L and $\phi_j = (j-1) \frac{\pi}{(L-1)}$ for j=1,..., 2(L-1). The arrays $a1_{m,n}$, $b1_{m,n}$, $a2_{m,n}$, $b2_{m,n}$, $v_{i,j}$, and $w_{i,j}$ are stored in A1(M,N), B1(M,N), A2(M,N), B2(M,N), V(I,J), and W(I,J), respectively. IDWV is the first dimension of the two-dimensional arrays V and W. MDAB is the first dimension of the two-dimensional arrays A1, B1, A2, and B2. WORK is a work array.

b) SUBROUTINE VSHS (L,IDVW,V,W,MDAB,A1,B1,A2,B2,WORK)

This subroutine computes the vector spherical harmonic synthesis. Given the coefficients $a1_{m,n}$, $b1_{m,n}$, $a2_{m,n}$, and $b2_{m,n}$ as input, subroutine VSHS computes $v_{i,j}$ and $w_{i,j}$ using (3.13).

4. THE VECTOR SPHERICAL HARMONIC ANALYSIS OF THE LINEARIZED SHALLOW-WATER EQUATIONS

A third package of subroutines is available for computing the modes of the linearized shallow-water equations. These equations describe the motion of a thin layer of incompressible fluid on the surface of a rotating sphere. These modes are of fundamental importance to meteorology and oceanography. If the equations are linearized with respect to the basic rotating state, then the resulting equations are called the Laplace tidal equations. Here we present an outline of their solution since the details are described in [10].

In this section we select meteorological notation in which ϕ and λ correspond to latitude and longitude, respectively. We also let u and v be the eastward and northward components of velocity and h be the radial displacement of the free surface from the mean height h_0. Using this notation the equations that describe the motion are

$$\frac{\partial u}{\partial t} - 2\Omega \sin\phi\, v = -\frac{g}{a\cos\phi}\frac{\partial h}{\partial \lambda} \tag{4.1}$$

$$\frac{\partial v}{\partial t} + 2\Omega \sin\phi\, u = -\frac{g}{a}\frac{\partial h}{\partial \phi} \tag{4.2}$$

$$\frac{\partial h}{\partial t} + \frac{h_o}{a\cos\phi}\left[\frac{\partial u}{\partial \lambda} + \frac{\partial}{\partial \phi}(v\cos\phi)\right] = 0 \tag{4.3}$$

where a and Ω are the radius and angular velocity of the earth, respectively, and g is the acceleration of gravity. If we introduce the dimensionless variables

$$\tilde{u} = \frac{u}{\sqrt{gh_o}}, \quad \tilde{v} = \frac{v}{\sqrt{gh_o}}, \quad \tilde{h} = \frac{h}{h_o}, \text{ and } \tilde{t} = 2\Omega t \tag{4.4}$$

then the equations (4.1) through (4.3) can be written

$$\frac{\partial \mathbf{W}}{\partial t} + \mathbf{LW} = 0 \tag{4.5}$$

where $\mathbf{W} = (\tilde{u}, \tilde{v}, \tilde{h})^T$ and $\mathbf{L}$ is the linear differential matrix operator.

$$\mathbf{L} = \begin{bmatrix} 0 & -\sin\phi & \dfrac{\gamma}{\cos\phi}\dfrac{\partial}{\partial\lambda} \\ \sin\phi & 0 & \gamma\dfrac{\partial}{\partial\phi} \\ \dfrac{\gamma}{\cos\phi}\dfrac{\partial}{\partial\lambda} & \dfrac{\gamma}{\cos\phi}\dfrac{\partial}{\partial\phi}[\cos\phi(\ )] & 0 \end{bmatrix} \tag{4.6}$$

in which

$$\gamma = \frac{\sqrt{gh_o}}{2a\Omega} \tag{4.7}$$

is a single dimensionless constant that characterizes the nature of shallow-water flows. A related quantity $\varepsilon = \gamma^{-2}$ is called Lamb's parameter. If we seek solutions in the form

$$\mathbf{W} = \mathbf{H}(\lambda,\phi)\; e^{-i\sigma\tilde{t}} \tag{4.8}$$

then from (4.5) the problem becomes one of determining eigenfrequencies σ and corresponding eigenfunctions (normal modes) $\mathbf{H}(\lambda,\phi)$ such that

$$(\mathbf{L} - i\sigma)\mathbf{H}(\lambda,\phi) = 0. \tag{4.9}$$

Let $\mathbf{U}$ and $\mathbf{V}$ be arbitrary vector functions and define the inner product

$$(\mathbf{U},\mathbf{V}) = \int_0^{2\pi} \int_{-\pi/2}^{\pi/2} \mathbf{U}^* \, \mathbf{V} \cos\phi \, d\phi \, d\lambda$$

then for real γ the vector differential operator $\mathbf{L}$ is antisymmetric, that is

$$(\mathbf{U},\mathbf{LV}) = -\,(\mathbf{LU},\mathbf{V}). \tag{4.10}$$

Thus the eigenfrequencies σ are real and the modes $\mathbf{H}(\lambda,\phi)$ that correspond to distant frequencies σ are orthogonal.

The goal is to determine $\mathbf{H}(\lambda,\phi)$ as an expansion in terms of the vector spherical harmonics. To this end we define three vector functions $\mathbf{Y}^m_{n,1}$, $\mathbf{Y}^m_{n,2}$, and $\mathbf{Y}^m_{n,3}$ that correspond to the vector spherical harmonics $B_{m,n}$, $C_{m,n}$, and $\mathbf{P}_{m,n}$, respectively, except for the order of the components and the scaling.

$$\mathbf{Y}_{n,1} = \begin{bmatrix} \dfrac{im}{\cos\phi}\,\bar{P}^m_n \\[2ex] \dfrac{d\bar{P}^m_n}{\partial\phi} \\ 0 \end{bmatrix} \frac{e^{im\lambda}}{\sqrt{n(n+1)}},$$

$$\mathbf{Y}_{n,2}^{m} = \begin{bmatrix} -\dfrac{d\bar{P}_n^m}{d\phi} \\ \dfrac{im}{\cos\phi}\bar{P}_n^m \\ 0 \end{bmatrix} \frac{e^{im\lambda}}{\sqrt{n(n+1)}} \tag{4.11}$$

$$\mathbf{Y}_{n,3}^{m} = \begin{bmatrix} 0 \\ 0 \\ \bar{P}_n^m \end{bmatrix} e^{im\lambda} .$$

The $\bar{P}_n^m$ are the normalized associated Legendre functions

$$\bar{P}_n^m = \sqrt{\frac{2n+1}{2}\,\frac{(n-m)!}{(n+m)!}}\; P_n^m . \tag{4.12}$$

We now seek eigenfrequencies σ and coefficients A_n^m, B_n^m, and C_n^m such that

$$\mathbf{H}(\lambda,\phi) = \sum_n (i\, A_n^m\, \mathbf{Y}_{n,1}^m + B_n^m\, \mathbf{Y}_{n,2}^m - C_n^m\, \mathbf{Y}_{n,3}^m) \tag{4.13}$$

and σ satisfy (4.9). If we substitute (4.13) into (4.9) it is nontrivial to show [10] that A_n^m, B_n^m, C_n^m, and σ must satisfy

$$(\sigma + q_n^m)\, A_n^m = r_n\, C_n^m + p_n^m\, B_{n-1}^m + p_{n+1}^m\, B_{n+1}^m \tag{4.14}$$

$$(\sigma + q_n^m)\, B_n^m = p_n^m\, A_{n-1}^m + p_{n+1}^m\, A_{n+1}^m \tag{4.15}$$

$$\sigma\, C_n^m = r_n\, A_n^m \tag{4.16}$$

where

$$p_n^m = \sqrt{\frac{(n-1)(n+1)(n-m)(n+m)}{n^2(2n-1)(2n+1)}}, \quad q_n^m = \frac{m}{n(n+1)} \tag{4.17}$$

and

$$r_n = \gamma \sqrt{n(n+1)}.$$

Thus the problem reduces to finding the eigenvalues σ and corresponding eigenvectors A_n^m, B_n^m, and C_n^m of an infinite matrix. The system (4.13) through (4.15) divides into two symmetric pentadiagonal subsystems. The first subsystem consists of the equations with unknowns A_n^m, B_{n+1}^m, and C_n^m for $n=m, m+2, m+4, \ldots$. The second subsystem consists of equations with unknowns A_{n+1}^m, B_n^m, and C_{n+1}^m for $n=m, m+2, m+4, \ldots$.

For $m > 0$ it is observed that the eigenvalues are distinct and hence the resulting eigenvectors and modes $\mathbf{H}(\lambda, \phi)$ are orthogonal. However, for the zonal case, $m=0$, $\sigma=0$ is a multiple eigenvalue and the corresponding eigenvectors are not necessarily orthogonal. Earlier, Kasahara [4] developed an orthogonal set of modes using the Gram-Schmidt process. More recently Kasahara and Swarztrauber followed a suggestion by Shigehisa [7] and defined the modes for m=0 as the limit of the modes as m goes to zero. This approach is attractive from the point of view that the resulting modes are orthogonal and exhibit the same fundamental physical characteristics that are displayed by the nonzonal modes. For example, the energy in the mode tends to concentrate at the equator as rotation is increased.

Near m=0 we assume that the coefficients have the following asymptotic forms

$$A_n^m = A_n m + O(m^2) \tag{4.18}$$

$$B_n^m = B_n + O(m) \tag{4.19}$$

$$C_n^m = C_n + 0(m) \tag{4.20}$$

$$\sigma = \sigma_1 m + 0(m^2) \tag{4.21}$$

If these forms are substituted into (4.14) through (4.16), then we find

$$r_n C_n + p_n^0 B_{n-1} + p_{n+1}^0 B_{n+1} = 0 \tag{4.22}$$

$$[\sigma_1 + \frac{1}{n(n+1)}]B_n = p_n^0 A_{n-1} + p_{n+1}^0 A_{n+1} \tag{4.23}$$

$$\sigma_1 C_n = r_n A_n. \tag{4.24}$$

These equations do not correspond to a standard eigenproblem, since σ does not appear in (4.22). However, if we eliminate A_n and C_n and define

$$\tilde{\sigma} = -\frac{1}{\sigma_1} \tag{4.25}$$

$$\tilde{B}_n = \frac{1}{\sqrt{n(N+1)}} B_n \tag{4.26}$$

$$d_n = \sqrt{n(n-1)}\,\frac{p_n^0}{r_n} \tag{4.27}$$

$$e_n = \sqrt{(n+1)(n+2)}\,\frac{p_{n+1}^0}{r_n} \tag{4.28}$$

then we obtain

$$e_{n-1}\, d_{n-1}\, \tilde{B}_{n-2} + [n(n+1) + e_{n-1}^2 + d_{n+1}^2 - \tilde{\sigma}]\, \tilde{B}_n$$
$$+ d_{n+1}\, e_{n+1}\, \tilde{B}_{n+2} = 0 \tag{4.29}$$

which corresponds to a standard eigenproblem. From this equation we obtain two independent tridiagonal systems. The first consists of the equations with unknowns $\tilde{B}_n$ for n = -1, 1,3, . . . and the second with n = 0,2,4, ... Once the eigenproblem is solved, then the results can be retraced in order to compute A_n, B_n, and C_n, which in turn determines the modes $\mathbf{H}(\lambda,\phi)$ for the case $m = \sigma = 0$. In [10] we show that the resulting modes are orthogonal.

The software package for computing the frequencies and modes $\mathbf{H}(\lambda,\phi)$ consists of three main subroutines. The first subroutine, called SIGMA, computes a user specified number of frequencies σ corresponding to a zonal wave number m. The second subroutine, called ABCOEF, computes a single set of coefficients A_n^m, B_n^m, and C_n^m corresponding to one of the frequencies computed by SIGMA. The third subroutine, called UVH, uses the coefficients computed by ABCOEF to tabulate a vector function $\mathbf{H}(\phi)$ at user-specified values of latitude ϕ. The mode can then be tabulated from the relation

$$\mathbf{H}(\lambda,\phi) = \mathbf{H}(\phi)e^{im\lambda}. \tag{4.30}$$

The modes $\mathbf{H}(\lambda,\phi)$ are called the Hough vector harmonics and $\mathbf{H}(\phi)$ is called the Hough vector function after Hough [3], who first investigated these modes.

a) SUBROUTINE SIGMA (M,MAXL,IERR,EPS,EASTGS,WESTGS, ROTATS,W)

This subroutine computes the frequencies σ as eigenvalues of (4.14) through (4.16) or $\tilde{\sigma}$ as eigenvalues of (4.29), depending on whether m is nonzero or zero, respectively. Input parameter EPS is Lamb's parameter ε. The eigenvalues are ordered and then partitioned into three groups. They are classified as westward, rotational, or eastward gravity depending on whether they occur in the lowest, middle, or highest third, respectively. They are computed and returned in the arrays WESTGS, ROTATS, and EASTGS, each of which contain the user-specified number of frequencies MAXL.

b) SUBROUTINE ABCOEF(M,MAXL,L,IEWR,IERR,SIG,EPS,BETA,A,B,C,W)

This subroutine computes the coefficients A_n^m, B_n^m, and C_n^m in the expansion of the mode $\mathbf{H}(\lambda,\phi)$. These coefficients are stored in the arrays A, B, and C, respectively. The input parameter L is the meridional index of the frequency SIG. As an example, if the user is interested in computing the Lth eastward gravity mode, then the program that calls ABCOEF should contain the statement SIG = EASTGS (L+1). Furthermore, IEWR should be set to 1 since IEWR takes the values 1, 2, or 3 depending on whether the desired mode is eastward gravity, westward gravity, or rotational. The parameter BETA is used to compute the Haurwitz modes which are solutions of (4.1) - (4.3) but not Hough functions, see [10].

c) SUBROUTINE UVH (M,MAXL,L,IEWR,EPS,NT,PHI,A,B,C,U,V,H,W)

This subroutine tabulates the components u, v, and h of the Hough vector function $\mathbf{H}(\phi)$ at the user-specified latitudes ϕ_j for j=1,..., NT which are stored in the array PHI. The input arrays are the same as the arrays A, B, and C that were computed by subroutine ABCOEF.

5. UTILITY SOFTWARE PACKAGES

In this section we present two software packages that are used by the software that was presented in the previous sections. The first is ALFPACK, which contains 16 subroutine user entry points and about 3,000 lines of code for computing single and double precision normalized associated Legendre functions $\bar{P}_n^m(\theta)$. The second package is FFTPACK, which is used to transform the longitudinal direction in the programs that analyze functions on the sphere. FFTPACK contains 19 subroutines and about 3,000 lines of code. The first lists the subroutines in ALFPACK and outlines their function. A complete description of their use is included with the package.

a) ALFPACK

i) SUBROUTINE LFK (INIT,N,M,CP,W)

The $P_n^m(\theta)$ can be computed as solutions of

$$\frac{1}{\sin\theta}\frac{d}{d\theta}\left(\sin\theta\,\frac{dP_n^m}{d\theta}\right) + \left[n(n+1) - \frac{m^2}{\sin^2\theta}\right]P_n^m = 0. \tag{5.1}$$

If we assume that P_n^m has the form (2.6) and substitute it into (5.1), then the coefficients $a_{m,n}(k) = b_k$ must satisfy

$$[(k-1)(k-2) - n(n+1)]\,b_{k-2} - 2[k^2 - n(n+1) + 2m^2]\,b_k$$
$$+ [(k+1)(k+2) - n(n+1)]\,b_{k+2} = 0. \tag{5.2}$$

Subroutine LFK computes the coefficients b_k and stores them in the array CP. The input parameter INIT is for initialization and W is a work array. These coefficients are used by the subroutine LFP to tabulate $\bar{P}_n^m$. They can also be useful for applications in which quantities are derived from $\bar{P}_n^m$, such as certain derivatives or integrals. Subroutines LFK and LFP can be used to compute $\bar{P}_n^m$ without computing $\bar{P}_n^m$ for any other values of m and n.

ii) SUBROUTINE LFP (INIT,N,M,L,CP,PB,W)

This subroutine uses the coefficients CP computed by LFK to tabulate $\bar{P}_n^m$ at the colatitudes $\theta_i = (i-1)\pi/(L-1)$ for $i=1, \ldots, L$. $\bar{P}_n^m$ is stored in the array PB.

iii) SUBROUTINE LFPT (N,M,THETA,CP,PB)

This subroutine computes a single value of $\bar{P}_n^m$ at θ = THETA, using the coefficients CP computed by LFK.

iv) SUBROUTINE LFMA (INIT,N,L,T,PB,W)

For fixed N and $\theta = (I-1)\pi/(L-1)$, this subroutine tabulates $\bar{P}_n^m$ for $m=0,\ldots, N$. They are computed as solutions of the three-term recurrence

$$\sin\theta\sqrt{(n-m)(n+m+1)}\,\bar{P}_n^{m+1} + 2m\cos\theta\,\bar{P}_n^m + \sin\theta\,\sqrt{(n-m+1)(n+m)}\,\bar{P}_n^{m-1} = 0. \tag{5.3}$$

This recurrence is stabilized by first computing $\bar{P}_n^n$, $\bar{P}_n^0$, and $\bar{P}_n^1$ and then computing the remaining $\bar{P}_n^m$ as the solution of a symmetric tridiagonal system of equations.

v) SUBROUTINE LFMB (INIT,N,L,I,PB,W)

This subroutine is the same as LFMA except for the following: a call with INIT = 0 initializes the work array. A call of LFMB with INIT = 0 is necessary only if I or L is changed from a previous call. A call of LFMA with INIT = 0 is necessary only if N or L is changed from a previous call. One uses either LFMA or LFMB depending on which requires the fewest calls with INIT = 0.

vi) SUBROUTINE LFNA (INIT,M,L,I,PB,W)

For fixed m and $\theta = (I-1)\pi/(L-1)$, this subroutine tabulates $\bar{P}_n^m$ for n=m, ..., L-1. They are computed as the solution of the three-term recurrence relations

$$\left[\frac{(n-m+1)(n+m+1)}{(2n+3)(2n+1)}\right]^{1/2} \bar{P}_{n+1}^{m} - \cos\theta\, \bar{P}_n^m$$

$$- \left[\frac{(n+m)(n-m)}{(2n+1)(2n-1)}\right]^{1/2} \bar{P}_{n+1}^{m} = 0. \qquad (5.4)$$

This recurrence is stabilized by first computing $\bar{P}_n^n$, $\bar{P}_n^{L-1}$, and $\bar{P}_n^L$ and then computing the remaining $\bar{P}_n^m$ as the solution of a symmetric tridiagonal system of equations.

vii) SUBROUTINE LFNB (INIT,M,L,I,PB,W)

This subroutine is the same as LFNA except for the following. A call of LFNB is required only if I or L is changed from a previous call. A call of LFNA is required only if m or L is changed from a previous call. One uses either LFNA or LFNB depending on which subroutine requires the fewest calls with INIT = 0.

viii) SUBROUTINE LFNC (INIT,M,L,THETA,PB,W)

This subroutine is the same as subroutine LFNA except that the user specifies the value of colatitude θ = THETA.

ix) SUBROUTINE BELSOV (L,N,I,PB,W)

For fixed N, L, and $\theta = (I-1)\pi/(L-1)$, this subroutine computes $\bar{P}_n^m$ for m=0, . . . , N. This subroutine must be called with increasing values of N. For example, if $\bar{P}_{10}^m$ is desired then BELSOV must be called 11 times with N=0, 1, 2, . . . , 10. The subroutine uses the diagonal recurrence due to S.L. Belousov [1].

$$\bar{P}_n^m = \left[\frac{(2n+1)(m+n-1)(m+n-3)}{(2n-3)(m+n)(m+n-2)}\right]^{1/2} \bar{P}_{n-2}^{m-2}$$

$$- \left[\frac{(2n+1)(m+n-1)(n-m+1)}{(2n-1)(m+n)(m+n-2)}\right]^{1/2} \cos\theta \, \bar{P}_{n-1}^{m-2} \tag{5.5}$$

$$+ \left[\frac{(2n+1)(n-m)}{(2n-1)(n+m)}\right]^{1/2} \cos\theta \, \bar{P}_{n-1}^{m} .$$

The remaining subroutines DLFK, DLFP, DLFPT, DLFMA, DLFMB, DLFNA and DLFNB are the double precision versions of LFK, LFP, LFPT, LFMP, LFMB, LFNA, and LFNB, respectively.

b) FFTPACK

All of the fast Fourier transforms in FFTPACK are based on an autosort algorithm which is credited to Stockham by Cochran et al. [2]. The package includes transforms for complex, real, and several symmetric sequences. There are no restrictions on the number of points N; however, the usual considerations apply, namely, that all the programs are more efficient when N is highly composite.

i) SUBROUTINE RFFTI (N,WSAVE)
ii) SUBROUTINE RFFTF (N,R,WSAVE)
iii) SUBROUTINE RFFTB (N,R,WSAVE)

Subroutines RFFTF and RFFTB compute the forward and backward transforms, respectively, of a real periodic sequence. Subroutine RFFTI initializes the work array WSAVE that is used by both RFFTF and RFFTB.

iv) SUBROUTINE EZFFTI (N,WSAVE)
v) SUBROUTINE EZFFTF (N,R,AZERO,A,B,WSAVE)
vi) SUBROUTINE EZFFTB (N,R,AZERO,A,B,WSAVE)

These subroutines are simplified versions of subroutines RFFTI, RFFTF, and RFFTB. Subroutines EZFFTF computes coefficients a_k and b_k such that the input sequence r_i has the form

$$r_i = a_0 + \sum_{k=1}^{n/2} (a_k \cos k(i-1)\frac{2\pi}{n} + b_k \sin k(i-1)\frac{2\pi}{n})$$

for i = 1, ..., n.

The sequences r_i, a_0, a_k, b_k are stored in R(I), AZERO, A(K), and B(K), respectively. Given the coefficients a_k and b_k, these subroutines EZFFTB compute the sequence r_i. Subroutine EZFFTI initializes the WSAVE array that is used by both EZFFTF and EZFFTB.

vii) SUBROUTINE SINTI (N,WSAVE)
viii) SUBROUTINE SINT (N,X,WSAVE)

These subroutines compute the sine transforms of the sequence X_k. Given the input sequence X_k, SINT computes

$$X_i = 2 \sum_{k=1}^{n} X_k \sin ik \frac{\pi}{(n+1)}$$

where equals is interpreted in the FORTRAN or replacement context. Subroutine SINT is its own inverse to within a scale factor of 2(n+1).

ix) SUBROUTINE COSTI (N,WSAVE)
x) SUBROUTINE COST (N,X,WSAVE)

These subroutines compute the cosine transforms of the sequence X_k. Given the input sequence X_k, COST computes

$$X_i = X_1 + \sum_{k=2}^{n-1} X_k \cos [(k-1)(i-1)\frac{\pi}{n-1}] + (-1)^{(i-1)} X_n .$$

Subroutine COST is its own inverse to within a scale factor of 2(n-1).

xi) SUBROUTINE SINQI (N,WSAVE)
xii) SUBROUTINE SINQF (N,X,WSAVE)
xiii) SUBROUTINE SINQB (N,X,WSAVE)

These subroutines compute the quarter-wave sine transforms. Given the input sequence X_k, subroutine SINQB computes

$$X_i = 4 \sum_{k=1}^{n} X_k \sin\,[(2k-1)i\frac{\pi}{2n}\ .$$

Subroutine SINQF is the inverse of subroutine SINQB to within a scale factor of 4n.

xiv) SUBROUTINE COSQI (N,WSAVE)
xv) SUBROUTINE COSQF (N,X,WSAVE)
xvi) SUBROUTINE COSQB (N,X,WSAVE)

These subroutines compute the quarter-wave cosine transforms. Given the input sequence X_k, subroutine COSQB computes

$$X_i = 4 \sum_{k=1}^{n} X_k \cos\,[(2k-1)(i-1)\frac{\pi}{2n}]\ .$$

xvii) SUBROUTINE CFFTI (N,WSAVE)
xviii) SUBROUTINE CFFTF (N,C,WSAVE)
xix) SUBROUTINE CFFTB (N,C,WSAVE)

These subroutines compute the complex Fourier transforms. Given the complex input sequence C_k, subroutine CFFTB computes

$$c_j = \sum_{k=1}^{n} c_k \, e^{i(j-1)(k-1)\frac{2\pi}{n}}$$

Subroutine CFFTF is the inverse of CFFTB to within a scale factor of n.

REFERENCES

1. S.L. Belousov, Tables of Normalized Associated Legendre Polynomials, Pergamon Press, New York, 1962.

2. W.T. Cochran, **et al.**, "What is the Fast Fourier Transform?", IEEE Trans. Audio Electroacoust., 15 (1967), pp. 45-55.

3. S.S. Hough, "On the application of harmonic analysis to the dynamical theory of the tides: Part II. On the general integration of Laplace's dynamical equations", Phil. Trans. R. Soc., London, A191 (1898), pp. 139-185.

4. A. Kasahara, "Normal modes of ultralong waves in the atmosphere", Monthly Weather Review, 104(1976), pp. 669-690.

5. B. Machenhauer and R. Daley, A baroclinic primitive equation model with a spectral representation in three dimensions, Report No. 4, Institute for Theoretical Meteorology, Copenhagen University, 1972.

6. P.M. Morse and H. Feshback, Methods of Theoretical Physics, McGraw-Hill, New York, 1953.

7. Y. Shigehisa, "Normal modes of the shallow-water equations for zonal wave number zero", submitted to: Journal of Meteorological Society of Japan, 1983.

8. P.N. Swarztrauber, "On the spectral approximation of discrete scalar and vector functions on the sphere", SIAM J. Numer. Anal., 16 (1979), pp. 934-949.

9. P.N. Swarztrauber, "The approximation of vector functions and their derivatives on the sphere", SIAM J. Numer. Anal., 18 (1981), pp. 191-210.

10. P.N. Swarztrauber and A. Kasahara, "The vector harmonic analysis of the Laplace tidal equations", submitted to SIAM J. on Sci. and Stat. Comp., 1983.

Sponsored by the National Science Foundation.

National Center for Atmospheric Research
Boulder, Colorado 80307

DESIGN, DEVELOPMENT, AND USE OF THE FINITE ELEMENT MACHINE

Loyce Adams and Robert G. Voigt

INTRODUCTION

During the summer of 1976 a weekly seminar was held at ICASE to study developments in parallel computing. The regular participants were Richard Brice, Griffith Hamlin, Harry Jordan, John Knight, David Loendorf, Jerry Tucker, and Robert Voigt with managerial support provided by James Ortega (ICASE) and Robert Fulton (NASA). Prior to that time David Loendorf had begun to investigate ways to speed up the solution of structural analysis problems by introducing parallelism into the finite element process utilizing microprocessor technology. It was therefore natural that the group used problems in structural analysis as a focal point for discussions.

This emphasis on an application area was unique. At that time only two parallel systems were under development: the Illiac IV eventually installed at the NASA Ames Research Center and the C.mmp at Carnegie-Mellon University. Both of these systems were essentially general purpose devices; the Illiac IV was to be used for a variety of large scale scientific problems and the C.mmp was primarily a vehicle for research into a variety of computer science issues arising in parallelism. The group was interested in how an application area might drive a design and whether such a narrow focus might lead to major simplifications in both hardware and software. The influence of the application will be discussed further in the next section.

ISBN 0-12-546080-5

Another central theme of the discussions was the role of microprocessors. At the time such devices were in their infancies. Simple eight-bit processors were readily available but the more powerful sixteen-bit versions were not. Nevertheless, it was clear that microprocessors were going to grow rapidly in capability, and it was reasonable to consider what could be accomplished by developing a system out of many such devices.

Thus, the activity of the group focused on ways to utilize microprocessors in a system for solving problems in structural analysis via the finite element method. The ideas and concepts developed were organized into an initial hardware design done by Harry Jordan and reported in Jordan [1978]. The eventual manifestation of the design is known as the Finite Element Machine (FEM) and is discussed more thoroughly in Section 3.

The FEM has had a long development period and the way the machine is to be used has undergone numerous changes. Some of the reasons for the extended development time are discussed in Section 4. Section 5 contains a brief discussion of the type of results that have been obtained using FEM and the paper concludes with some observations about developing research computers. Finally, the bibliography contains all work that has been published on FEM as of this writing. Many of these papers are not cited but are included here for completeness.

2. MACHINE DESIGN ISSUES

We will now discuss some of the issues considered by the research group which influenced the design of FEM. In its simplest form the finite element method for the case of static stress analysis may be described as follows:

1. subdivide the region of interest into elements,
2. choose basis functions spanning the space in which the approximate solution lies,
3. integrate the basis functions over each element to determine its contribution,

4. assemble the contributions of all the elements into a single system

$$K x = f \tag{2.1}$$

5. solve (2.1) for the approximate solution x.

For more details the reader is referred to the finite element literature, for example, Strang and Fix [1973].

When the above process is implemented on a serial computer the majority of the time is consumed by steps 3 and 5. In addition certain solution techniques for Eq. (2.1) do not require the actual formation of the stiffness matrix K. Thus the activity of the research group focused on steps 3 and 5.

In order to have a focal point for discussion consider the simplified planar structure in Figure 1.

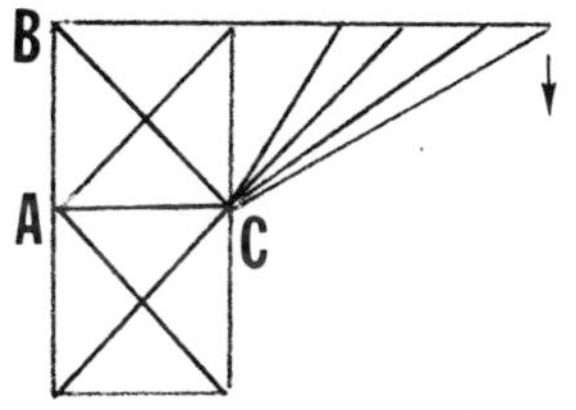

Figure 1. Example Structure

Assume we are interested in determining the stresses in the structure if a force is applied as indicated by the arrow. Further assume that the structure is modeled by different elements such as beams and plates and that an appropriate set of basis functions has been chosen. Then from step 3 the basis functions must be integrated over each element. These integrations may be done in parallel for each element. However since the elements may be different or since similar elements may have different material properties, it is not possible to execute the same instruction sequence across all of the elements. Thus in order to achieve the maximum degree of parallelism it was considered desirable for FEM to be a parallel system of multiple-instruction-multiple-data (MIMD) type in the classification of Flynn [1966].

For the solution of Eq. (2.1) both direct and iterative methods were considered. For most applications of interest the matrix K is symmetric, positive definite and banded with bandwidth β as indicated in Figure 2.

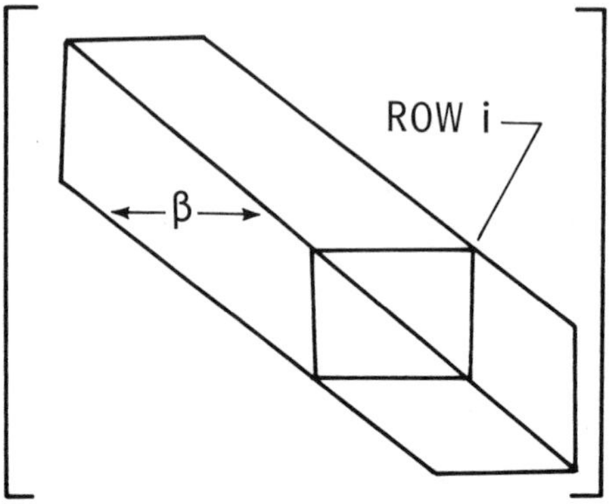

Figure 2. Form of the Stiffness Matrix

In a direct method such as Cholesky factorization, at the ith step conceptually rows $i + 1$ through $i + \beta$ are modified using the pivot row i. It is possible for these modifications to be done in parallel; however, row $i + \beta + 1$ can not be modified until computation on row $i + 1$ has been completed. Thus the degree of parallelism in the sense of Hockney and Jesshope [1981] is limited to β unless one is prepared to consider parallelism at the operation level within each row. The latter is possible but raises serious questions about interprocessor communication for the element in the ith column of the pivot row must be made available to all processors containing elements of the ith column that are due to be modified by the pivot row.

Finally there is the usual problem of fill associated with direct methods. In general all elements within the band will become non-zero during the factorization. This destroys the sparsity of the matrix and greatly increases the storage requirements.

Iterative methods do not suffer from the fill associated with direct methods. In addition it is easier to obtain a higher degree of parallelism. For example, if we consider the iterative method

$$x^{k+1} = Bx^k + d, \tag{2.2}$$

where x^k represents the approximate solution vector, the degree of parallelism is N, the number of nodes of the discretization. This leads to the concept of a node per processor.

In addition to the increased parallelism this approach also offers the advantage of requiring primarily only local communication. Writing equation (2.2) as

$$x_i^{k+1} = \sum_{j \in I_i} b_{ij} x_j^k + d_i,$$

we see that x_i^{k+1} depends only on a relatively small number of values of x^k as indicated by the index set I_i which consists of those nodes which are physically connected to node x_i. Thus it is desirable to have communication paths between the processor containing x_i and the other processors containing x_j for all $j \in I_i$. Therefore it was decided that each processor should be connected to its eight nearest neighbors in the plane so as to support the communication required by triangles, an important part of many real structures, see Figure 3.

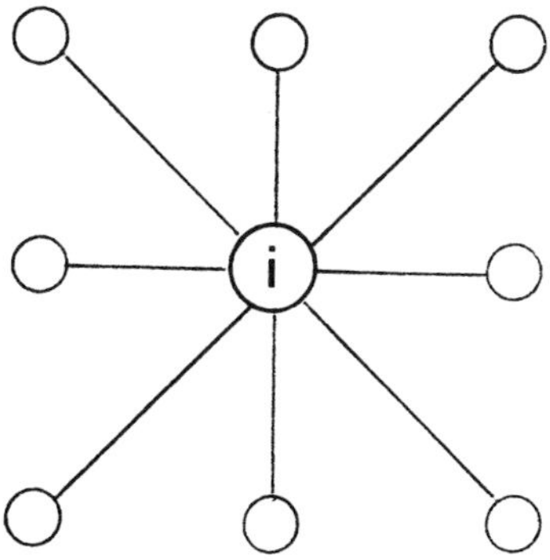

Figure 3. Eight nearest neighbor communication paths for processor i.

It should be noted that the connectivity sets I_i do not all represent the same pattern or number of connections. For example contrast the connectivity of nodes A and B in Figure 1. This means that the computations required for updating each node will not be the same and hence reinforces the requirement that the system be of MIMD type.

Equation (2.2) is the prototype of the classical Jacobi iteration which exhibits the maximum degree of parallelism but does not have as desirable convergence characteristics as methods like Gauss-Seidel. In Gauss-Seidel like methods, x_i^{k+1} depends on other values at the (k+1) step and thus was not thought of as a parallel method. However, many authors have pointed out that Gauss-Seidel can be turned into a parallel method by employing the so-called red-black or checkerboard ordering, see for example, Ortega and Voigt [1977]. Thus the eight nearest neighbor connection would support the use of modern iterative methods on FEM.

A significant problem remains: one must be able to map the discretized structure of interest onto the processors of the FEM so that all nodes that are connected lie on processors that are connected. This turns out to be a nontrivial problem even if the degree of connectivity of every node is eight or less, see for example, Bokhari [1979]. However many structures contain nodes that are connected to more than eight other nodes as node C in Figure 1. Communication required by such nodes obviously cannot be supported directly by the eight nearest neighbor connections. Thus it was decided to augment the so-called local processor connections with a global bus which provides a connection between any two processors. The work of Bokhari focused on finding mappings of the nodes onto the processors so as to minimize the use of the global bus which was viewed as a resource that could be easily saturated.

At this point the design appeared to hold considerable promise for the classical iterative methods, but there was also interest in studying the more modern accelerations of these methods, as well as the conjugate gradient method and its many variants. A key step in these methods requires parameters which are obtained by computing inner products involving the approximate residual and direction vectors. In the scenario described above the approximate solution is distributed across the processors and it requires O(n) steps to accumulate an inner product using local connections on an n×n array. To overcome this delay a separate circuit was designed that connects the processors in a classical binary tree. This made it possible to find the maximum element of a vector or to sum the elements of a vector in O(log n) time when the elements were distributed across the n×n array. For additional details see Jordan et al. [1979].

At this point the basic concepts of the FEM were fixed and a preliminary design for 1024 processors was done by Jordan (see Jordan [1978]) under support from the Structures Division at the NASA Langley Research Center (LaRC). In 1979 LaRC began fabrication of an experimental system under the leadership of David Loendorf with hardware integration support provided by Frank Mewszel; in 1981 David Loendorf left Langley and Olaf Storaasli assumed responsibility for system development. The prototype presently contains eight processors with expansion continuing; a 36 processor version is shown in Figure 4. The system is discussed in more detail in the next section.

3. THE CURRENT FINITE ELEMENT MACHINE

In this section we describe the hardware and system software for both the controller and the nodal processors. More detailed descriptions of the hardware may be found in Jordan [1978], Jordan, et al. [1979], and Loendorf [1983]. A summary of the current system software may be found in Storassli, et al. [1982] and detailed descriptions of the controller support software and array software may be found in Knott [1983a] and Crockett [1983] respectively.

Hardware

The controller consists of a TI 990/10 minicomputer with 128K words of memory, four 5-megabyte disk drives, a Kennedy 9000 tape drive, and a line printer. The purposes of the controller are to serve as the user interface to the FEM array by providing program development and problem definition

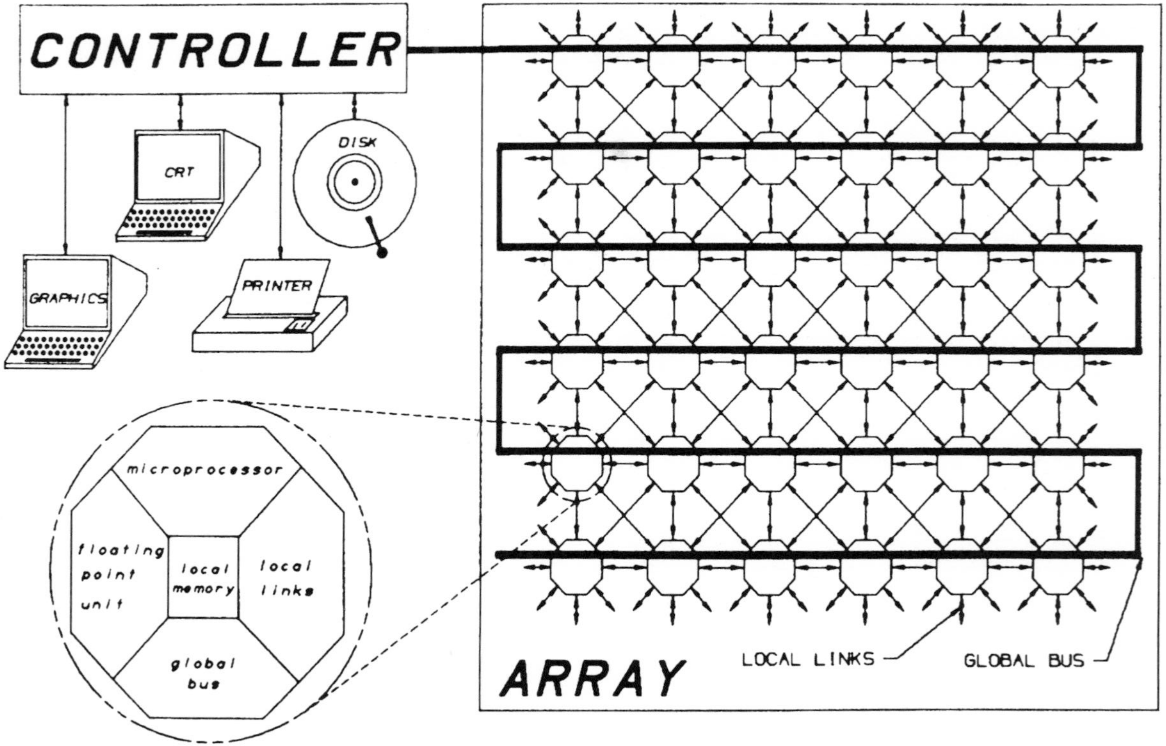

Figure 4. Finite Element Machine System

tools, to provide mass storage for programs and nodal processor input and output data, and to initiate and monitor activity on the array.

Each nodal processor in the array is comprised of three hardware boards: the CPU board, the IO-1 board, and the IO-2 board. The CPU board contains a Texas Instrument 9900 16-bit microprocessor, 16K bytes of erasable programmable read only memory (EPROM), 32K bytes of random access memory (RAM), and an Advanced Micro Devices AM9512 floating point chip. The EPROM and 4K of RAM are reserved for system software. The remaining 28K RAM is available for program code, run-time data structures, and input data. The AM9512 floating point chip with a clock frequency of 2 MHz provides single precision (32-bit, 25-bit mantissa) and double precision (64-bit, 57-bit mantissa) add, subtract, multiply, and divide operations. To use this capability, the operands must be loaded by the nodal processors' system software which requires approximately 360 microseconds for two single precision numbers (this number was obtained through private discussions with Tom Crockett). Once the operands are loaded, a single precision floating point multiply can be performed in approximately 100 microseconds.

The IO-1 board contains twelve local communication links and the summation/maximum hardware. Each link is a 1.5 MHz bit serial interface with an associated hardware FIFO buffer capable of storing 16 16-bit words of input data from a neighboring processor. The links are normally configured in an eight nearest neighbor with torroidal wrap around scheme but may be changed before each program execution to support other strategies. Likewise, an output register holds values that have been transferred from the memory on the CPU board for transmission to neighboring processor(s). The summation/maximum hardware allows p values, one per processor, to be added in $\log_2 p$ time by providing a binary tree structure with the processors initially at the leaves of the tree. This hardware works independent of the other communication net works of the machine and was designed specifically to perform summations (needed by inner products) and determine maximum values (desirable for norm calculations).

The IO-2 board contains the global bus connections, the signal flag networks, and the processor's self-identification tag. The global bus is a 1.25 MHz time-multiplexed 16-bit parallel bus that connects all processors to each other and to the controller. The bus has hardware FIFO buffers on both the input and output lines capable of storing 64 words of data for buffering purposes. The bus serves as the vehicle for transmitting the program code and data from the controller disk to each nodal processor. The bus is also used during the execution of application programs to transmit data between non-neighboring processors with each processor having equal priority for the bus on a first-come, first-serve basis, see Knott and Crockett [1982]. The bus can be used in the broadcast mode to send information to a set of processors from another processor or the controller. The signal flag hardware connects a processor to eight separate binary flag networks which span all processors. Any or all of these hardware flags can be enabled or disabled during program execution and allow for synchronization and decision making. A processor's physical self-identification number is hard wired on the IO-2 board and is matched to the logical processor number of a particular application code by the system software for use in interprocessor communication and decision making.

At present, eight processors, all connected via local links to each other, have been running application codes. Currently, another eight processor system is being installed providing one system for hardware and any additional software development and one for application users. The next step will be to add eight processors to one system for a 4×4 FEM array. Eventually a larger array may be built if studies performed on the 4×4 array indicate that such an effort is warranted.

<u>System Software</u>

The system software consists of the vendor's standard software for the TI 990 controller, the FEM Array Control Software (FACS) that runs on the TI 990 and provides the user interface to the array, the NODAL executive operating system that runs on each TI 9900 microprocessor in the array, and the PASCAL Library extensions (PASLIB) that support access to the architectural features of the array like communication and synchronization. A short description of each of these software components and how they work together to implement applications programs follow.

The vendor's software for the TI 990 controller includes a screen editor, an assembler, a reverse assembler, a Pascal compiler, and a link editor. The applications programmer uses this software to edit, compile, and link his program to be run on the array. Typically this program will be executed by all the processors in the array with different data. The programmer can use an interactive graphics interface or the text editor to model his problem and partition this data to separate data files (stored on the controller) for each processor in the array. Alternatively, a heuristic utility program may be written to partition the data for the processors, Bokhari [1979].

After the program and data files for each processor have been stored on the 990 disk, the FACS software is used in conjunction with the nodal EXEC operating system on each of the nodal processors to initialize the array, select the array configuration, define the size of the data areas (memory on the nodal processors that contains either initial data or intermediate data between job runs), load any or all of these data areas from the data files on the controller, and download (broadcast) the program linked code. All these FEM commands are implemented as control language procedures in FACS which is a natural extension of the menu-driven command interpreter of the vendor software. The programmer must therefore create a command program which describes which program(s) and data are to be down loaded and the appropriate sequence for that downloading and execution. This command program in turn is invoked by a single controller command. After the program begins execution on the array, the controller enters an interactive execute mode and receives all messages/errors from all array processors but displays on the user's terminal information from only one preselected processor. During execution, the FACS software maintains a file of all output data received from the array, errors encountered by all the processors, and a log of the events during the job run which can be post processed by utility programs at the end of the job session. FACS also provides interactive debugging commands that allow the user to single step, halt, kill, resume, dump memory, set program breakpoints, and inspect and change memory, status, and registers.

The two components of the system software that run on each TI 9900 microprocessor are the NODAL EXEC operating system and the PASLIB routines. NODAL EXEC is stored in EPROM on each TI 9900 and provides interrupt handling, basic I/O, timing, memory allocation, task management, and a command monitor. In addition, NODAL EXEC contains a package of command routines which implement all functions the Controller commands the TI 9900 to perform. Typical functions include loading object code, loading data into the data areas, establishing processor connectivity (local and global neighbors), executing programs, performing debugging, and uploading results.

Perhaps of more interest to the applications programmer are the PASLIB routines. PASLIB is a library of Pascal subroutines that allow the programmer to use the local links, global bus, signal flag network, sum/max circuit, and the AM9512 floating point unit as well as communicate with the controller. The most commonly used routines are written in assembly language and stored in EPROM. A few of these will now be described.

To synchronize using flag i, processors must first call the ENABLE (flag i) routine to add this flag to the network, after which a call to the BARRIER (flag i) routine will cause all processors with flag i enabled to synchronize. Note that these routines must be called by all processors wishing to synchronize. The BARRIER routine may be used in iterative algorithms to synchronize before a call to the ALL (flag i) routine is executed to check for global convergence.

To send n words of data that are stored in memory starting at location ℓ to processor p the programmer would call the PASLIB routine SEND(p,ℓ,n) or SEND2(p,ℓ,i,n) if data is distinguished by an index tag i. Data may also be broadcast to all local and global neighbors by SENDALL(ℓ,n) or SENDAL2(ℓ,i,n).

Data is received from another processor in either a synchronous or asynchronous mode (which has to be defined by the programmer in the command file on the controller). For the synchronous mode, input from the sending processor is queued in the order it is received and must be read by the receiver in this order. For the asynchronous mode, only the most recently received record (for each index tag) is saved. By providing these two modes of communication, the system software must necessarily be more general and therefore more expensive; however, they provide a mechanism for studying both synchronous and asynchronous algorithms.

To use the AM9512 floating point unit, the operands must be loaded via PASLIB routines. For example, to multiply two numbers x and y and store in z, the appropriate statement would be z := MULT(x,y). This adds a cost of 358 μs to execute the MULT procedure compared to the 99 μs for actually performing the multiply on the AM9512. (This tremendous overhead is due to the incompatibility of the AM9512 and the TI 9900 that could not be avoided at the time the hardware selections were made.)

4. FEM DEVELOPMENT EXPERIENCES

FEM development to date has provided a number of learning experiences which may be useful to share. Progress has been slower than anticipated for a variety of reasons involving both hardware and software issues.

At the time the microprocessor was selected, the TI 9900 was the only 16-bit processor available. As development progressed a number of unexpected small hardware purchases were required. Significant delays in procurement were encountered due both to delay in manufacturer delivery and to federal procurement policies. In hindsight the low-bidder competitive procurement process of the government was often not the most effective strategy to purchase small quantities of scarce parts to meet the requirements of an evolving research system. Any cost benefits from competitive procurements were negated by delays in system development incurred while waiting for deliveries. A better strategy might have been a master contract for all parts with the specifics to be determined as work progressed.

The design itself required the usual modifications but a serious weakness was the omission of any hardware error detection. The latter situation caused significant delays in the debugging process that was already complicated by the presence of several processors functioning independently.

As with most research projects funding was limited and staffing levels were barely adequate to encompass the hardware, software, numerical analysis and applications disciplines. Furthermore when some initial hardware became operational, it was difficult to satisfy the needs of both those doing hardware enhancement and those doing systems/applications development -- activities equally important for such research projects. The competiton for access was finally resolved by establishing a dual system, presently consisting of eight processors for hardware development and another eight for software.

Not surprisingly there were also difficulties in the software development. The original idea of choosing a controller with the same instruction set as the processors in the array and thereby using that software as a basis for the array software seemed sound. However, the software underwent such signficant changes that it might have been better to develop an all new software system. Major issues revolved around the adaptation of Pascal to the operating environment of the array. For example floating point arithmetic had to be adapted to account for the presence of the AM9512. This involved facilities for moving data to that device as well as converting the data to the appropriate format.

Perhaps the biggest issue was to provide support for the variety of communication mechanisms available. Since this is a major issue in any parallel computing system we will discuss it in more detail below. It should be noted first however that the fact that the FEM software provides an effective environment for the user is a credit to the efforts of Tom Crockett and Judson Knott. Their task was further complicated by changes in the way users expected to utilize the array, an example of which follows.

As discussed earlier, the original concept of the FEM involved considering a node, or possibly an element, of the discretization per processor. The implementation of a standard iterative method would then require frequent communication in which a processor would send a value to each of its neighbors and receive a value from each. As it turned out such frequent bursts of communication involving only a few words of data were inefficient.

This inefficiency can be better understood by considering the steps a processor must complete to actually send data:

1. interrupt the processor,
2. copy the data to be sent to an output buffer,
3. place the output buffer in a queue for either local or global transmission,
4. generate a send interrupt,
5. execute the send interrupt,
6. transmit the data.

Executing these steps in order to send two words of data requires approximately 1.8 milliseconds with the actual transmission contributing only a few microseconds. When one compares this time with the approximately .46 milliseconds required by a floating point operation, one is led to try to organize the computation so that the frequency of transmission is decreased while the amount of data per transmission is increased.

One way to accomplish this change in transmission style is to place several nodes or elements in each processor. In fact, note that if a two-dimensional region is stored in each processor then for many iterative methods only the boundary data must be exchanged between the processors. If the region is $q \times q$ then the amount of data transmitted is $O(4q)$ while the computation per processor is $O(q^2)$. This provides a mechanism for balancing the communication time with the computation time. However the array is now being used differently than was originally intended further complicating the software development.

5. USE OF THE FEM

The results obtained on FEM so far can be put into two separate categories; namely, parameter results and numerical algorithm results especially appropriate for static stress analysis.

Tom Crockett and Judd Knott have timed the system software routines that send and receive data and perform synchronization using the flag network. These parameters are given in Adams [1982] and were used there as input to a model to predict performance of some numerical algorithms as the number of processors increased. Smith and Loendorf [1982] have also obtained FEM parameters for use in performance evaluation. For the most recent version of the operating system software, Knott [1983b] has timed the PASLIB routines that send and receive a package of numbers between neighbor processors as a function of the package size. The results of his study have given guidance to application programmers as to the tradeoffs in package size selection.

Execution times of several iterative algorithms for fixed sized problems with the number of processors varying from one to five have been obtained. These times were used to compare the algorithms as a function of the number of processors and to obtain speedup results for a given algorithm. The iterative algorithms include multi-color SOR as described in Adams and Ortega [1982] with results reported in Adams [1982], conjugate gradient (Adams [1982]), and m-step preconditioned conjugate gradient (Adams [1983a] and [1983b]). In addition, execution time and speedup results for assembling the finite element stiffness matrix K of (2.1) in parallel is given in Adams [1982].

Additional studies are underway to determine execution time and speed up results for a variety of algorithms and applications. Direct methods for solving the system (2.1) are being compared particularly for multiple right hand sides. Direct and iterative methods for tridiagonal systems are also being compared. This problem is particularly important as a building block for other more general iterative methods.

A new application under study involves computing the nonlinear dynamic response of a structure under a prescribed load condition. The technique is highly parallel and preliminary results indicate that excellent speedups are possible.

6. CONCLUSIONS

The FEM is important as a model of the kind of research system that is required to understand the issues in parallel computing and to evaluate various techniques for making such systems useful. While the development of the FEM took longer than anticiapted, a great deal has been learned during that process. A collection of researchers have been able to evaluate techniques for programming MIMD systems. They have developed an understanding of how to use communication among the processors and have actually measured the performance of that communication in solving real problems. They are beginning to learn just how hard it can be to debug a program for a system of asynchronous processors. This kind of experience may lead to improved techniques for tracking and isolating errors in MIMD systems.

There is no scarcity of ideas in the computer science community about how one should design an MIMD system. There is a scarcity of systems on which to evaluate these ideas. Such systems are absolutely essential if we are to develop the knowledge required to produce useful MIMD systems. Despite some shortcomings, the FEM is an excellent example of how important such a test bed can be.

7. ACKNOWLEDGEMENT

The authors are indebted to Robert Fulton for many useful discussions and for suggestions for improving the manuscript.

REFERENCES

[1] Adams, G., D. Mullens, and A. Shah [1979], "Software Design Project: Finite Element System Executive," Report No. CSGD-79-1, Computer Systems Design Group, Univeristy of Colorado.

[2] Adams, L. M. [1982], "Iterative Algorithms for Large Sparse Linear Systems on Parallel Computers, Ph.D. Thesis, University of Viriginia; also published as NASA CR-166027, NASA Langley Research Center, November.

[3] Adams, L. M. [1983a], "M-Step Preconditioned Conjugate Gradient Methods," ICASE Report No. 83-9, NASA Langley Research Center, Hampton, VA, April.

[4] Adams, L. M. [1983b], "An M-Step Preconditioned Conjugate Gradient Method for Parallel Computation," Proc. of the 1983 Intl. Conference on Parallel Processing, IEEE Catalog No. 83CH1922-4, August, pp. 36-43.

[5] Adams, L. M. and J. M. Ortega [1982], "A Multi-Color SOR Method for Parallel Computation," Proc. of the 1982 Intl. Conference on Parallel Processing, IEEE Catalog No. 82CH1794-7, August, pp. 53-56.

[6] Bokhari, S. H. [1979], "On the Mapping Problem for the Finite Element Machine, Proc. of the 1979 Intl. Conference on Parallel Processing, IEEE Catalog No. 79CH1433-2C, August, pp. 239-248.

[7] Bokhari, S. H. [1981], "On the Mapping Problem," IEEE Trans. Computers, Vol. C-30, March, pp. 207-214.

[8] Bokhari, S. H. [1981], "MAX: An Algorithm for Finding Maximum in an Array Processor with a Global Bus," Proc. of the 1981 Intl. Conference on Parallel Processing, IEEE Catalog No. 81CH1634-5, August, pp. 302-303.

[9] Bokhari, S. H. and A. D. Raza [1983], "Reducing the Diameters of Array," Report No. EECE-83-01, Department of Electrical Engineering, University of Engineering and Technology, Lahore, Pakistan, June.

[10] Bostic, S. W. [1983], "TEKLIB Graphics Library," NASA TM-84633, NASA Langley Research Center, Hampton, VA, March.

[11] Crockett, T. W. [1983], PASLIB Programmer's Guide for the Finite Element Machine, preliminary draft, to appear as NASA Contractor Report, NASA Langley Research Center, Hampton, VA.

[12] Chughtai, M. Ashraf [1982], "Complete Binary Spanning Trees of the Eight Nearest Neighbor Array," Report No. EECE-82-01, Department of Electrical Engineering, University of Engineering and Technology, Lahore, Pakistan, November.

[13] Gannon, D. [1980], "A Note on Pipelining a Mesh Connected Multiprocesor for Finite Element Problems by Nested Dissection," Proc. of the 1980 Intl. Conference on Parallel Processing, IEEE Catalog No. 80CH1569-3, August, pp. 197-204.

[14] Hockney, R. W. and C. R. Jesshope [1981], "Parallel Computers," Adam Hilger Ltd., Bristol, Great Britain.

[15] Iqbal, M. Ashraf and S. H. Bokhari [1983], "New Heuristics for the Mapping Problem," Report No. EECE-83-02, Department of Electrical Engineering, University of Engineering and Technology, Lahore, Pakistan, June.

[16] Jordan, H. F. [1978], "A Special Purpose Architecture for Finite Element Analysis," Proc. of the 1978 Intl. Conference on Parallel Processing, IEEE Catalog No. 78CH1321-9C, August, pp. 263-266.

[17] Jordan, H. F., ed. [1979], "The Finite Element Machine Programmer's Reference Manual," Report No. CSDG-79-2, Computer Systems Design Group, University of Colorado, Boulder, CO.

[18] Jordan, H. F. and D. A. Podsiadlo [1980], "A Conjugate Gradient Program for the Finite Element Machine," Computer Systems Design Group, University of Colorado, Boulder, CO.

[19] Jordan, H. F. and D. A. Podsiadlo [1981], "Operating Systems Support for the Finite Element Machine," Report No. CSDG-81-2, Computer Systems Design Group, University of Colorado, Boulder, CO.

[20] Jordan, H. F. and P. L. Sawyer [1978], "A Multi-Microprocessor System for Finite Element Structural

Analysis," Trends in Computerized Structural Analysis and Synthesis, A. K. Noor and H. G. McComb, Jr., eds., Pergamon Press, Oxford, pp. 21-29.

[21] Jordan, H. F., M. Scalabrin, and W. Calvert [1979], "A Comparison of Three Types of Multiprocessor Algorithms," Proc. of the 1979 Intl. Conference on Parallel Processing, IEEE Catalog No. 79CH1433-2C, August, pp. 231-238.

[22] Knott, J. D. [1983a], "FEM Array Control Software User's Guide," NASA CR 172189, NASA Langley Research Center, Hampton, VA.

[23] Knott, J. D. [1983b], "A Performance Analysis of the PASLIB Version 2.1X SEND and RECV Routines on the Finite Element Machine," NASA CR 172205, NASA Langley Research Center, Hampton, VA.

[24] Knott, J. D. and T. W. Crockett [1982], "Fair Dynamic Arbitration for a Multiprocessor Communication Bus," Computer Architecture News, Vol. 10, No. 5, September, pp. 4-9.

[25] Loendorf, D. [1983], "Advanced Computer Architecture for Engineering Analysis and Design," Ph.D. Thesis, University of Michigan, Aerospace Engineering Department.

[26] McBride, W. E. [1980], "Simulations Solution Algorithms for the FEM," Modeling and Simulation, Vol. 11, Part 2, Proc. of the 11th Annual Pittsburgh Conference on Modeling and Simulation, pp. 595-599.

[27] McBride, W. E. [1981], "Simulating the PAM (Purely Asynchronous Method) on the FEM (Finite Element Machine)," Modeling and Simulation, Vol. 12, Proc. of the 12th Annual Pittsburgh Conference on Modeling and Simulation, pp. 413-416.

[28] Mehrotra, P. and T. W. Pratt [1982], "Language Concepts for Distributed Processing of Large Arrays," ACM SIGACT-SIGOPS Symposium on Principles of Distributed Computing, August, pp. 19-28.

[29] Mehrotra, P. [1982], "Parallel Computation on Large Arrays," Ph.D. Thesis, University of Virginia.

[30] Ortega, J. and R. Voigt [1977], "Solutions of Partial Differential Equations on Vector Computers," Proc. 1977 Army Numerical Analysis Conference, pp. 475-526.

[31] Podsiadlo, D. A. [1981], "An Operating System for the Finite Element Machine," Report No. CSDS-81-3, Computer Systems Design Group, University of Colorado, Boulder, CO.

[32] Rea, G. C. [1983], "A Software Debugging Aid for the Finite Element Machine," Report No. CSDG-83-2, Computer Systems Design Group, University of Colorado, Boulder, CO.

[33] Shah, A. K. [1980], "'Group Broadcast' Mode of Interprocessor Communication for the Finite Element Machine," Report No. CSDG-80-1, Computer Systems Design Group, University of Colorado, Boulder, CO.

[34] Smith, C. and D. D. Loendorf [1982], "Performance Analysis of Software for an MIMD Computer," Performance Evaluation Review, Vol. 11, No. 4, Proc. of the 1982 ACM Sigmetrics Conference on Measurement and Modeling of Computer Systems, pp. 151-162.

[35] Storaasli, O. O., S. W. Peebles, T. W. Crockett, J. D. Knott, and L. M. Adams [1982], "The Finite Element Machine: An Experiment in Parallel Processing," NASA TM-84514, NASA Langley Research Center, July.

[36] Voitus, R. F. [1981], "MPSP: A Multiple Process Software Package for the Finite Element Machine," Report No. CSDG-81-4, Computer Systems Design Group, University of Colorado, Boulder, CO.

Supported by NASA under NASA Contract Nos. NAS1-17070 and NAS1-17130 while the authors were in residence at ICASE.

Institute for Computer Applications in Science and Engineering, NASA Langley Research Center, Hampton, VA 23665

INDEX

W

Z